A Comprehensive Database of Tests on Axially Loaded Piles Driven in Sand

A Comprehensive Database of Tests on Axially Loaded Piles Driven in Sand

Zhongxuan Yang
Zhejiang University, Hangzhou, China

Richard Jardine
Imperial College, London, UK

Wangbo Guo
Zhejiang University, Hangzhou, China

Fiona Chow
Woodside Energy Ltd, Perth, Australia

AMSTERDAM • BOSTON • HEIDELBERG • LONDON
NEW YORK • OXFORD • PARIS • SAN DIEGO
SAN FRANCISCO • SINGAPORE • SYDNEY • TOKYO
Academic Press is an imprint of Elsevier

Academic Press is an imprint of Elsevier
125 London Wall, London EC2Y 5AS, UK
525 B Street, Suite 1800, San Diego, CA 92101-4495, USA
225 Wyman Street, Waltham, MA 02451, USA
The Boulevard, Langford Lane, Kidlington, Oxford OX5 1GB, UK

Notices
Knowledge and best practice in this field are constantly changing. As new research and experience broaden our understanding, changes in research methods, professional practices, or medical treatment may become necessary.

Practitioners and researchers must always rely on their own experience and knowledge in evaluating and using any information, methods, compounds, or experiments described herein. In using such information or methods they should be mindful of their own safety and the safety of others, including parties for whom they have a professional responsibility.

To the fullest extent of the law, neither the Publisher nor the authors, contributors, or editors, assume any liability for any injury and/or damage to persons or property as a matter of products liability, negligence or otherwise, or from any use or operation of any methods, products, instructions, or ideas contained in the material herein.

ISBN: 978-0-12-804655-5

British Library Cataloguing-in-Publication Data
A catalogue record for this book is available from the British Library

Library of Congress Cataloging-in-Publication Data
A catalog record for this book is available from the Library of Congress

For Information on all Academic Press publications
visit our website at http://store.elsevier.com/

Publisher: Joe Hayton
Acquisition Editor: Simon Tian
Editorial Project Manager: Naomi Robertson
Production Project Manager: Julie-Ann Stansfield
Designer: Victoria Pearson Esser

Typeset by TNQ Books and Journals
www.tnq.co.in

Printed and bound in the United States of America

Contents

Preface

Driven piles are commonly used to support structures such as major bridges, offshore platforms, wind turbines, and even domestic housing. However, practical tools for calculating their capacity have, until recently, been relatively unreliable, especially in sand and soft rock ground conditions. Research at Imperial College London led to new calculation methods, as summarized by Jardine and Chow (1996) and Jardine et al. (2005) for both clays and sands, with the latter involving a "CPT-based" procedure. Several other groups proposed alternative "CPT-based" sand approaches in 2005. Research over the following decade has included investigations into the effects of soil layering, sand mineralogy, pile age, cyclic loading, and lateral response. Simplified "CPT-based sand" axial calculation methods have also been tabled and included in industrial design guidance; alternative procedures have been proposed for clays and other geomaterials.

Database verification plays a vital role in checking the predictive reliability of both existing and new methods of capacity assessment. Historical database assessments have shown the relatively poor reliability of the conventional procedures applied most commonly in practical design, identifying that sand and soft rock ground conditions leads to the greatest degrees of scatter and bias. While the four CPT-based sand procedures cited by API GEO (2014) are all thought to lead to better reliability parameters than their conventional "Main Text" method, it is recognized that the statistical outcomes depend on the characteristics of the pile load test databases. The test population size, pile dimension, and ground condition ranges are important, as are the qualities of the tests and associated site investigations. The two most substantial databases related to CPT-based sand methods are those assembled by Imperial College (Chow, 1997; Jardine et al., 2005) and the University of Western Australia (Lehane et al., 2005a). Taken together, these comprise just over 100 tests that cover a wide range of pile types and ground conditions. However, some test entries have been questioned on grounds of test quality or data documentation. A reexamination of the available information appears to be necessary.

The scope for moving forwards toward internationally agreed new design procedures depends on updating, upgrading, and augmenting the pile test databases, along with gaining experience on their practical application. The need for improved pile load test databases led to a joint Zhejiang University (ZJU)—Imperial College London (ICL) project starting in mid-2011, mainly supported by grants from Ministry of Education and Natural Science Foundation of China, and Ministry of Science and Technology of China.

This booklet comprises the first major output from the study. It presents an updated database for piles driven in predominantly silica sands, drawing in new data entries from the team's own projects, the literature and from acknowledged communication with other research groups worldwide. A detailed description of the background and analytical approach is offered along with an appendix that details each of the assembled tests and its site conditions. Yang et al. (2015b) provide a summary of the approach taken, the preliminary conclusions drawn by the ZJU-ICL team and give details of how our database may be accessed freely to enable other research groups to test and evaluate their design methods independently.

The ZJU-ICL team will continue their work over the next 3 years by updating their electronic sand test database regularly and producing further outputs that will both cover axial capacities for piles driven in other geomaterials and updated guidance on load–displacement behavior.

Zhongxuan Yang
Richard Jardine
Wangbo Guo
Fiona Chow
September 2015

Acknowledgments

The Authors wish to acknowledge the many contributions made to the long-term research by current and former colleagues at Imperial College and ZJU. They also thank the many research groups who have provided the data, particularly Prof Rücker from Federal Institute for Materials Research and Testin in Germany, Dr Tsuha from University of São Paulo at São Carlos in Brazil, Prof. Mayne and Dr Niazi from Georgia Tech in USA, and Prof. Frank and Dr Burlon from IFSTTAR in France. The ZJU-ICL database project is funded by the National Key Basic Research Program of China (No. 2015CB057801), International Science and Technology Cooperation Program of China (No. 2015DFE72830), the Natural Science Foundation of China (Grant Numbers: 51178421, 51322809, 51578499), and the Chinese Ministry of Education Distinguished Overseas Professorship Programme, and the Fundamental Research Fund for Central Universities (No. 2014XZZX003-15).

Notation List

A_b Pile base area

A_r Area ratio

$A_{rs,eff}$ Effective area ratio

B Outer width for square pile

D Pile diameter

d_{50} Mean particle size

D_{CPT} Diameter of standard CPT devices = 0.036 m

D_i Pile inner diameter

D_r Sand relative density

h Relative height above the tip

K_f Lateral effective earth pressure coefficient

L Embedded pile length

ΔL_p Increment of soil plug

N_q Bearing capacity factor

p_a Absolute atmospheric pressure = 100 kPa

$q_{b,0.1}$ End bearing available after displacement by 10% of the pile diameter

$q_{b,lim}$ Limiting unit end-bearing capacity

q_c CPT tip resistance

Q_{c1N} Normalized cone tip resistance

$q_{c,avg}$ Average cone tip resistance around pile tip

$q_{c,tip}$ Cone tip resistance at pile tip

q_t CPT end resistance with pore pressure correction for piezocones

Q_b Sum of the base capacity

Q_c Calculated pile axial capacity

Q_m Capacities measured in careful field tests

Q_s Sum of the shaft capacity

Q_{total} Total capacity of pile

R Pile radius

R_a Interface roughness

R_{CLA}	Center line average roughness
R^*	Equivalent pile radius
μ	Statistical mean value
s	Standard deviation
t	Pile wall thickness for open-ended pile
z	Refer to any given depth below ground level
Δz	Increment of the pile displacement
z_{tip}	Pile tip depth
β	Shaft friction factor
γ	In situ bulk unit weight of soils
δ_f	Interface shearing angle
σ'_{v0}	Free-field vertical effective stress
σ'_{rc}	Radial stresses acting on pile shaft
$\Delta\sigma'_{rd}$	Change in shaft radial stress during loading
τ_f	Local ultimate shaft friction
$\tau_{f,lim}$	Limiting shaft friction
$\tau_{f,avg}$	Averaged external skin friction
API	American Petroleum Institute
CoV	Coefficient of variation
CPT	Cone penetration test
FFR	Final filling ratio
ICP	Imperial college pile
IFR	Incremental filling ratio
LCPC	Laboratoire Central des Ponts et Chaussées (France)
NGI	Norwegian Geotechnical Institute
PSD	Particle size distribution
UWA	University of Western Australia
ZJU-ICL	Zhejiang University and Imperial College London

Chapter | One

Introduction

This booklet provides an updated database research resource of high-quality load tests relating to piles driven in predominantly silica sands. It comprises the first major output from the Zhejiang University/Imperial College London (ZJU-ICL) pile database study that commenced in mid-2011 and offers a detailed description of the project's background and analytical approach before providing a substantial appendix that details the tests and their site conditions. A preliminary evaluation is also given of how several design methods perform when tested against the new database, and conclusions are drawn regarding the methods' practical application and the effects of pile age after driving. The ZJU-ICL team's work will continue: first by updating the associated electronic database for piles driven in silica sands as new tests become available; secondly by considering both different geomaterials; and thirdly by addressing the piles' load–displacement responses.

Predicting the behavior of piles driven in sand is an important industrial question, particularly in major bridge, harbor, and offshore engineering applications, where it can affect the practicality and economics of major projects. Axial capacity predictions are crucial to, for example, tension leg, tripod, or jacket offshore structures; see for example, Overy (2007), Merritt et al. (2012), or Jardine (2013). Foundation stiffness can also be important to structural fatigue life and the operations of facilities such as wind turbines. Database studies in which field measurements are compared with predictive calculations show that the accuracies and reliabilities of the industry's routine design tools are often far lower than practitioners appreciate. Briaud and Tucker (1988) demonstrated that conventional axial capacity calculations (Q_c) show considerable bias and scatter when compared to the capacities (Q_m) measured in careful field tests. Their statistical analysis of Q_c/Q_m ratios quantified the reliability of a range of design methods, reporting mean values μ and coefficients of variations (CoV—defined as the standard deviation, s, divided by μ). Ideally, the mean Q_c/Q_m should be close to unity and the CoV (or s) as low as possible.

Fundamental research with field ICPs (Imperial College Piles) equipped with high-quality surface stress transducers (SSTs) by Lehane et al. (1993), and Chow (1997) revealed that the routine design methods fail to capture key aspects of the stress regime that develops around pile tips and shafts during penetration in sand. The tip stresses were found to correlate directly with local cone penetration test (CPT) q_c resistances, as did the radial stresses, σ'_{rc}, set up on the pile shafts. The latter also reduced systematically, at any given depth (z) below ground level, as the pile tip advanced and the relative height above the tip, $h = z - z_{tip}$, increased. A weak dependence on the free-field vertical effective stress, σ'_{v0}, was also identified. Jardine and Chow (1996) gave functions that related σ'_{rc} to q_c, σ'_{v0}, and h/R for closed-ended piles that only required slight modification (substituting an equivalent radius R^*) to be used for open-ended piles. The ICP experiments also showed that shaft loading generated local radial stress changes that varied with pile diameter and loading sense, while local shaft failure developed once a critical state interface shear angle δ_f was mobilized that could be predicted from laboratory tests and correlated with grain size and pile shaft roughness. The simple expressions proposed were able to capture the key physical phenomena that control field shaft capacity and end-bearing capacity trends.

Subsequent research has considered additional factors, including the influence of the load cycles imposed during installation (White and Lehane 2004; Jardine et al., 2013a), time effects (Jardine et al., 2006; Gavin et al., 2013; Karlsrud et al., 2014), how particle breakage under the tip and surface abrasion affect the stresses and develop a well-defined interface shear zone (Yang et al., 2010), the stress regime developed in the surrounding soil mass (Jardine et al., 2009, 2013b), and the influence of cyclic loading (Jardine et al., 2005; Tsuha et al., 2012). Yang et al. (2014) and Zhang et al. (2014) have gone on to relate the stress measurements made with experimental investigations and numerical analyses by other workers.

New practical design tools were proposed from the work of Lehane et al. (1993) and Chow (1997), which evolved into the updated Imperial College (ICP-05) method detailed by Jardine et al. (2005). Other groups developed alternative approaches that recognized similar features of physical behavior through alternative formulations. These include the Fugro-05 (Kolk et al., 2005a), Norwegian Geotechnical Institute (NGI-05, Clausen et al., 2005), and University of Western Australia (UWA-05, Lehane et al., 2005a). Rigorous database studies have become key tools to assess the potential efficacy of these new design procedures. Lehane and Jardine (1994), Chow (1997), Kolk et al. (2005a), Clausen et al. (2005), Jardine et al. (2005), and Schneider et al. (2008) all assembled databases to test their new CPT-based design procedures in comparison with the offshore industry-standard "Main Text" API RP2GEO (2014) approach and its forebears. They found that the "Main Text" approach was subject to surprisingly high overall CoVs in Q_c/Q_m (up to 0.88) when predicting compression capacity in sand and that the new procedures led to lower CoVs and less bias with respect to pile geometry (diameter D and L/D ratio, where L = embedded length), loading sense (tension or compression), and sand relative density (D_r). Williams et al. (1997), Jardine et al. (2005) reported case histories where the "Main Text" API approach gave Q_c/Q_m values ranging from 0.4 to 2.9.

The current API RP2GEO (2014) acknowledges the limitations of its Main Text approach and the potential advantages of the new CPT methods. But it also notes industry's lack of experience with the new methods. Practitioners are uncertain as to which, if any, of the four methods they should adopt for routine application and their assessments are made difficult by the general limitations of the pile test databases, and in particular a lack of high-quality tests on large open-ended piles in silica sands at sites that have been characterized to a high standard.

The most comprehensive sets appear to be those assembled by Lehane and Jardine (1994), Chow (1997), and Jardine et al. (2005), termed herein as the ICP database, and that published by Lehane et al. (2005a) and Schneider et al. (2008), which we term the UWA database. Taken together, these include over 100 different piles driven in silica sand and tested to failure. However, only 11 piles (from just three sites) were open-ended, had $D \geq 600$ mm and full CPT profiles. Further tests are required to augment this sparse data set, obtain information from a wider range of international sites and gain further insight into uncertain factors such as the effects of layering on base resistance (Xu, 2006) and the effects of pile age on capacity (Jardine et al., 2006; Karlsrud et al., 2014 or Gavin et al., 2013).

The ZJU-ICL project initiated in 2011 has now gathered an augmented database of high-quality pile load tests that should allow closer testing and perhaps development of alternative design procedures for piles driven in predominantly silica sands. The following pages outline the background to the work, the methodology adopted, the population of current entries, and the digital reporting format before describing some preliminary results obtained in comparisons between axial capacities calculated by various approaches and the site measurements. Yang et al. (2015b) outline how the web-based electronic database may be accessed, used, and added to by other workers. The booklet includes a major appendix that provides all the key information relating to the database. It also illustrates the minimum pile test data requirements and sets out the reporting format.

We trust that the booklet, which can be considered as a highly detailed addendum to Yang et al. (2015b), will allow other researchers to test their pile design methods independently. As mentioned earlier, the ZJU-ICL project will continue by updating the web-based database for piles in silica sands as new tests become available. It will also move on to consider different geomaterials and axial pile load–displacement response.

Chapter | Two

Design Methods and Database Assessments

CHAPTER OUTLINE

2.1 CURRENT DESIGN METHODS FOR PILES DRIVEN IN SAND

The shaft resistances of piles driven in sand can often be mobilized fully at axial displacements smaller than 10% of the piles' diameters D. However, far larger displacements may be required to mobilize end bearing fully, especially with large open-ended piles. Such displacements cannot be tolerated in most cases, so the static axial compression capacity is often defined as the load Q_{total} that can be mobilized under a limiting displacement of $0.1\,D$. Q_{total} is then the sum of the shaft capacity Q_s and base capacity Q_b:

$$Q_{\text{total}} = Q_s + Q_b = \pi D \int \tau_f \, dz + q_{b,0.1} A_b \qquad (2.1)$$

where τ_f is the local ultimate shaft friction; z is pile depth; $q_{b,0.1}$ is the end bearing available after a displacement of $D/10$, and A_b is conventionally defined as the full pile base area. End bearing is considered negligible under tension loading.

The API "Main Text" method (API RP2GEO, 2014) and its forebears have long been the mainstream international design tool for large driven piles. The approach, which is outlined in Table 2.1, assumes that local shaft and base resistances grow in proportion with the free-field vertical effective stress, with $\tau_f = \beta\sigma'_{v0}$ and $q_b = N_q\sigma'_{v0}$. The rules do not recognize any relative pile tip depth dependency of shaft resistance, but specify upper limits to the unit shaft and base resistances ($\tau_{f,\text{lim}}$ and $q_{b,\text{lim}}$). The original API formulation presented the β term as $\beta = K_f \tan\delta_f$. With open-ended pile pipes that drove without plugging, the constant lateral earth pressure coefficient K_f was taken as 0.8 under both compression and tension loading. The K_f value for plugged pipe piles or closed-ended piles was assumed equal to unity, while interface shear angles δ_f were adopted that are now known to be physically improbable. The current method's updated key parameters (β, $\tau_{f,\text{lim}}$, and $q_{b,\text{lim}}$) are specified in Table 2.2.

A Comprehensive Database of Tests on Axially Loaded Piles Driven in Sand

TABLE 2.1 API Design Method (API RP2GEO, 2014)

	Design equations	Notation
Shaft friction τ_f	$\tau_f = K_f \sigma'_{v0} \tan \delta_f = \beta \sigma'_{v0} \le \tau_{f,lim}$ API-RP2GEO (2014) combines K_f and δ_f as $\beta = K_f \tan \delta_f$. The values in Table 2.2 are applicable to open-ended piles; for closed-ended or fully plugged pipe piles, β values are considered to be 25% greater.	Coefficients β, N_q, $\tau_{f,lim}$, and $q_{b,lim}$ can be found from Table 2.2
Base resistance $q_{b,0.1}$	$q_{b,0.1} = N_q \sigma'_{v0} \le q_{b,lim}$	

TABLE 2.2 Design Parameters for Cohesionless Siliceous Soil

Relative density[a]	Soil description	Shaft friction factor[b] β (-)	Limiting shaft friction values $\tau_{f,lim}$ (kPa)	End-bearing factor N_q (-)	Limiting unit end-bearing values $q_{b,lim}$ (MPa)
Very loose	Sand	Not applicable[d]	Not applicable[d]	Not applicable[d]	Not applicable[d]
Loose	Sand				
Loose	Sand-silt[c]				
Medium dense	Silt				
Dense	Silt				
Medium dense	Sand-silt[c]	0.29	67	12	3
Medium dense	Sand	0.37	81	20	5
Dense	Sand-silt[c]				
Dense	Sand	0.46	96	40	10
Very dense	Sand-silt[c]				
Very dense	Sand	0.56	115	50	12

Note: The parameters listed in the table are intended as guidelines only. Where detailed information, such as CPT records, strength tests on high-quality samples, model tests, or pile driving performance, is available, other values may be justified.

[a] *The definitions for the relative density percentage are as follows:*

- *Very loose, 0–15;*
- *Loose, 15–35;*
- *Medium dense, 35–65;*
- *Dense, 65–85;*
- *Very dense, 85–100.*

[b] *The shaft friction factor β (equivalent to the "K tan δ_f" term used in previous editions of API 2A-WSD) is introduced in this edition to avoid confusion with the δ_f parameter used in the annex.*

[c] *Sand-silt includes those soils with significant fractions of both sand and silt. Strength values generally increase with increasing sand fractions and decrease with increasing silt fractions.*

[d] *Design parameters given in previous edition of API 2A-WSD for these soil/relative density combinations may be unconservative. Hence, it is recommended to use CPT-based methods from the annex for these soils.*

After API (2014).

API RP2GEO (2014) recognizes that the earlier Main Text approaches (e.g., API, 2000) were nonconservative for loose sands and consider the Main Text approach unsuitable for such cases. The database outlined in this booklet contains some loose sand cases and for these we apply the API (2000) guidance, giving in effect a hybrid of the 2014 and 2000 recommendations. We refer to these predictions as "API Main Text" predictions in the remainder of the booklet.

As noted earlier, the API Main Text approach suffers from significant bias and scatter. Other "conventional" approaches, such as French Road and Bridges (LCPC) cone penetration test (CPT) method (Bustamante and Gianeselli, 1982), can provide lower Q_c/Q_m CoVs (in the 0.4–0.5 range) for piles driven in sand. But they may also suffer from significant nonconservative bias when applied to large driven piles (Chow, 1997). More modern "CPT-based" design procedures that better match the physical processes that control axial pile behavior in sands include the updated ICP-05 (Jardine et al., 2005), UWA-05 (Lehane et al., 2005a), Fugro-05 (Kolk et al., 2005a), and NGI-05 (Clausen et al., 2005) methods. The API RP2GEO (2014) Commentary now cautiously cites "simplified" and "offshore" variants of the ICP and UWA methods along with the Fugro and NGI approaches. While the "CPT-based" methods have now been used in practice for some time (particularly the ICP and its "MTD" forebear), practitioners often apply the full ICP approach in place of the simplified procedure: see for example, Williams et al. (1997), Overy (2007), or Merritt et al. (2012).

The four CPT-based methods are applicable to both closed and open-ended piles. Their key starting point is recognition that pile driving affects the stress regime around the pile shaft and that the simple earth pressure model implicit in the API Main Text's $\tau_f = \beta\sigma'_{v0}$ approach is not reasonable physically. Following from the in situ instrumented pile tests by Lehane et al. (1993) and Chow (1997), the radial effective stress acting on the pile shaft at any particular depth, z, below ground surface is considered to be significantly affected by other factors, including the local sand state and CPT q_c value, the relative height of the point above the tip h/R, the free-field vertical effective stress σ'_{v0}, and the pile tip geometry (closed or open) and wall thickness if open. The local radial effective stress was written in the ICP as:

$$\sigma'_{rc} = f\left(q_c, \frac{h}{R^*}, \sigma'_{v0}\right) \tag{2.2}$$

where the equivalent pile radius $R^* = (R^2_{outer} - R^2_{inner})^{0.5}$ with open-ended piles and $R^* = R = D/2$ for closed-ended piles.

Lehane et al. (1993) and Chow (1997) showed that local shaft failure obeys the Coulomb failure criterion at the pile–soil interface and that the local ultimate shaft friction τ_f developed under static testing can be calculated by:

$$\tau_f = \sigma'_{rf} \tan\delta_f \tag{2.3}$$

where δ_f is the constant volume of the interface shearing angle that can be obtained from large displacement ring shear tests conducted as set out by Jardine et al. (2005). Such tests give values that are very different to those specified by API (2000). The local radial stress at failure σ'_{rf} is expected to differ from that resulting from installation σ'_{rc}, and the ICP-05 method specifies expressions that allow for both the difference between compression and tension loading and the effect of restrained interface dilation $\Delta\sigma'_{rd}$, which adds to the shaft friction by an amount that increases with sand shear stiffness (calculated from q_c and σ'_{v0}) and pile roughness, but diminishes with increasing pile radius R.

All of the expressions necessary to calculate ultimate ICP shaft friction values τ_f are given in Table 2.3, while Tables 2.4–2.6 outline the equivalent expressions employed in the UWA, Fugro, and NGI methods, respectively. To save confusion, these tables employ the same format as the API RP2GEO commentary sections rather than those in which they were presented by the original authors. The "simplified" ICP-05 form recommended by API removes the shaft loading dilatancy term $\Delta\sigma'_{rd}$ listed in Table 2.3. Other input parameters are rounded up or down.

The UWA-05 approach started from the ICP framework, adding a new "effective area" term to Eqn (2.2), removing the mild dependency on σ'_{v0} and relating shaft friction to h/D rather than h/R^*. The "offshore version" of UWA-05 proposed by Lehane et al. (2005a) neglects the shaft dilatancy term and assumes a fully coring installation mode when calculating the effective area term implicit in the shaft radial effective stress expression.

The Fugro-05 approach also built on the ICP (or MTD as it was known then) method. It effectively combined Eqns (2.2) and (2.3) after recalibrating the terms relating τ_f to q_c, h/R, σ'_{v0} through regression analysis involving Kolk et al.'s (2005b) pile load test database. Implicit in the Fugro method is that the interface

TABLE 2.3 ICP-05 Design Method (Jardine et al., 2005)

	Design equations	Notation
Shaft friction τ_f	Full version: $\tau_f = a\left[0.029bq_c\left(\frac{\sigma'_{v0}}{p_a}\right)^{0.13}\left[\max\left(\frac{h}{R^*},8\right)\right]^{-0.38}+\Delta\sigma'_{rd}\right]\tan\delta_f$ "Simplified" version: $\tau_f = a\left[0.023bq_c\left(\frac{\sigma'_{v0}}{p_a}\right)^{0.10}\left[\max\left(\frac{h}{R^*},8\right)\right]^{-0.40}\right]\tan\delta_f$ $a=0.9$ for open-ended piles in tension; 1.0 for other conditions. $b=0.8$ for piles in tension; 1.0 for piles in compression.	
Base resistance $q_{b,0.1}$	$q_{b,0.1}/q_{c,avg}=\max[1-0.5\log(D/D_{CPT}),0.3]$	Closed-ended
	if $D_i \geq 2.0(D_r-0.3)$ or $D_i \geq 0.083q_{c,avg}/p_a D_{CPT}$ (D_i is in meters), then the pile is "unplugged", and $q_{b,0.1}/q_{c,avg}=A_r$. On the contrary, the pile is "plugged" and $q_{b,0.1}/q_{c,avg}=\max[0.5-0.25\log(D/D_{CPT}),0.15,A_r]$	Open-ended

TABLE 2.4 UWA-05 Design Method (Lehane et al., 2005a)

	Design equations	Notation
Shaft friction τ_f	Full version: $\tau_f = \frac{f_t}{f_c}\left[0.03 q_c A_{rs,eff}^{0.3}\left[\max\left(\frac{h}{D},2\right)\right]^{-0.5} + \Delta\sigma'_{rd}\right]\tan\delta_f$ Offshore version: $\tau_f = \frac{f_t}{f_c}\left[0.03 q_c A_r^{0.3}\left[\max\left(\frac{h}{D},2\right)\right]^{-0.5}\right]\tan\delta_f$ Effective area ratio $A_{rs,eff} = 1 - \mathrm{IFR}(D_i/D)^2$ Area ratio $A_r = 1 - (D_i/D)^2$; IFR ($=\Delta L_p/\Delta z$), if IFR is not measured, take $\mathrm{IFR} = \min[1, (D_i/1.5)^{0.2}]$ Ratio of tension to compression capacity $f_t/f_c = 1.0$ for compression and 0.75 for tension	
Base resistance $q_{b,0.1}$	$q_{b,0.1} = 0.6 q_{c,avg}$	Closed-ended
	Full version: $q_{b,0.1}/q_{c,avg} = 0.15 + 0.45 A_{rb,eff}$ Offshore version: $q_{b,0.1}/q_{c,avg} = 0.15 + 0.45 A_r$ Effective area ratio: $A_{rb,eff} = 1.0 - \mathrm{FFR}(D_i/D)^2$ $\mathrm{FFR} \approx \min[1, (D_i(m)/1.5)^{0.2}]$	Open-ended

TABLE 2.5 Fugro-05 Design Method (Kolk et al., 2005a)

	Design equations	Notation
Shaft friction τ_f	$\tau_f = 0.08 q_c \left(\frac{\sigma'_{v0}}{p_a}\right)^{0.05}\left(\frac{h}{R^*}\right)^{-0.90}$	Compression with $h/R^* \geq 4$
	$\tau_f = 0.08 q_c \left(\frac{\sigma'_{v0}}{p_a}\right)^{0.05}(4)^{-0.90}\left(\frac{h}{4R^*}\right)$	Compression with $h/R^* \leq 4$
	$\tau_f = 0.045 q_c \left(\frac{\sigma'_{v0}}{p_a}\right)^{0.15}\left[\max\left(\frac{h}{R^*},4\right)\right]^{-0.85}$	Tension
Base resistance $q_{b,0.1}$	$q_{b,0.1} = 8.5(p_a q_{c,avg})^{0.5} A_r^{0.25}$	Open- and closed-ended

shear angle δ_f is fixed at 29° and does not vary with sand type. The NGI-05 approach starts from different principles and leads to a direct expression for the τ_f available at any given depth z that relies on the assessment of local relative density, rather than a direct link with q_c. Unlike the other three methods, it allows for the effect of relative pile tip depth (h) through a "sliding triangle" approach in which the reduction of local shaft resistance depends only on the relative depth z/L, where L is the final embedded shaft length. Here the "friction fatigue"

TABLE 2.6 NGI-05 Design Method (Clausen et al., 2005)

	Design equations	Notation
Shaft friction τ_f	$\tau_f = \max(z/L p_a F_{D_r} F_{sig} F_{tip} F_{load} F_{mat}, \tau_{min})$ $F_{D_r} = 2.1(D_r - 0.1)^{1.7}$, $D_r = 0.4\ln(q_{c1N}/22)$ $F_{sig} = (\sigma'_{v0}/p_a)^{0.25}$, $\tau_{min} = 0.1\sigma'_{v0}$ $F_{tip} = 1.0$ for open-ended and 1.6 for closed-ended $F_{load} = 1.0$ for tension and 1.3 for compression $F_{mat} = 1.0$ for steel and 1.2 for concrete	
Base resistance $q_{b,0.1}$	$q_{b,0.1} = F_{D_r} q_{c,tip} = 0.8 q_{c,tip}/(1+D_r^2)$	Closed-ended
	$q_{b,0.1} = \min[\text{plugged } q_{b,0.1}, \text{unplugged } q_{b,0.1}]$ plugged $q_{b,0.1} = F_{D_r} q_{c,tip} = 0.7 q_{c,tip}/(1+3D_r^2)$ unplugged $q_{b,0.1} = q_{b,ann} A_r + q_{b,plug}(1 - A_r)$ $q_{b,ann} = q_{c,tip}$, $q_{b,plug} = 12\tau_{f,avg} L/(\pi D_i)$	Open-ended

depends on neither the absolute values of h or D, nor the ratio h/D. The method employs other empirical factors to deal with pile material, end conditions, and loading type (tension or compression).

Different $q_{b,0.1}$ expressions are used in each method, which are also summarized in Tables 2.3–2.6. There is no difference between the full and "simplified" ICP base resistance expressions, but the "offshore UWA" method assumes full coring during driving as part of its base capacity calculation (Lehane et al., 2005b). The shaft and base UWA-05 expressions therefore differ between the "full" and "offshore" versions, as set out in Table 2.4.

A variety of techniques are also recommended for averaging the key input CPT tip design value $q_{c,avg}$ required for end-bearing assessment. Both ICP-05 and Fugro-05 employed the LCPC recommendation of averaging q_c over an interval $\pm 1.5D$ above and below the pile tip level, while UWA-05 adopted the "Dutch" averaging technique. NGI-05 employed the relative density D_r at the tip level itself to calculate a $q_{c,tip}$ input parameter through the given approximate correlation with σ'_{v0}. Research by Xu (2006) has shown that pile base resistance is more strongly affected than had been thought by any weak layers beneath and above the pile tip. The ICP-05 authors have proposed recently (see Yang et al., 2015b) that while the best estimate (average) profile may be used to estimate shaft resistance in design, a lower bound profile should be adopted for base capacity assessment.

2.2 DATABASE ASSESSMENTS OF METHOD UNCERTAINTY AND RELIABILITY

Rigorous database studies have become key tools to assess the potential efficacy of new design procedures such as those outlined above. Lehane and Jardine (1994), Chow (1997), Kolk et al. (2005a), Clausen et al. (2005), Jardine et al. (2005), and Schneider et al. (2008) all assembled databases to either test or calibrate their CPT-based design procedures and make comparisons with the offshore industry-standard "Main Text" API RP2GEO (2014) approach and its forebears.

The main features of the Jardine et al. (2005) "ICP" database are summarized in Table 2.7, while Table 2.8 gives the headline statistical Q_c/Q_m outcomes found for the ICP-05 and Main Text API methods, considering the mean μ, and ± Coefficients of variations (CoV) values for Q_c/Q_m. Figures 2.1–2.4 present the associated scatter diagrams of Q_c/Q_m against D_r and slenderness ratio L/D for the ICP

TABLE 2.7 General Characteristics of ICP, UWA, and ZJU-ICL Databases

	ICP database (2005)	UWA database	ZJU-ICL database
Total no of tests	64 + 19 = 83	77	116
New tests	83	26	48
ZJU-ICL accepted new tests[a]	54	65 (ICP = 51, UWA = 14)[b]	116 (ICP = 54, UWA = 14, ZJU-ICL = 48)
Pile types	Mainly driven piles, but with 8 jacked and 1 vibro-driven	Only driven piles	Only driven piles
Pile shape	Circular, square, and octagonal piles	Circular, square, and octagonal piles	Circular, square, and octagonal piles
Pile diameter (mm)	200–2000	200–2000	200–2000
Pile length (m)	5.3–46.7	5.3–79.1	5.3–79.1
Soil description	Mainly siliceous sands, carbonate contents less than 15%, shaft length in clay less than 40%	Pile tips bearing a siliceous sand and siliceous sand contributes >50% of shaft capacity	Pile tips bearing a siliceous sand and siliceous sand contributes >65% of shaft capacity
Load test	Static; base and shaft capacity separated individually	Static	Static
Failure criterion	If no clear peak indicated in compression, pile head displacement of 0.1 *D* (outer diameter);	If no clear peak indicated in compression, pile tip displacement of 0.1 *D* (outer diameter);	If no clear peak indicated in compression, pile head displacement of 0.1 *D* (outer diameter);
	Failure in tension was usually well defined	Tension was defined as the maximum uplift load minus pile weight	Tension was defined as the maximum uplift load minus pile weight
Age on testing	Pile tests were conducted 0.5–200 days after driving. Average age after driving = 34 days after driving. Time details were reported in 74% of case records	Time between driving and load testing is typically 0.5–200 days (average t = 24 days). Time details were reported in 77% of the case records	Pile tests were conducted 0.5–220 days, with an average of t = 33 days after driving. Time details were reported in 65% of the case records

[a] *The number of tests that pass the ZJU-ICL criteria and are included in the new database.*
[b] *The bracketed figures denote the number of tests accepted into the ZJU-ICL database from each specified source.*

TABLE 2.8 Assessment of Shaft Capacity Predictions: Q_c/Q_m

Method	Mean (μ)	Standard deviation (s)	Coefficient of variation (CoV = s/μ)
ICP, all piles	0.99	0.28	0.28
ICP, all open-ended piles	1.05	0.30	0.28
API RP2A (1993); all piles	0.87	0.58	0.60

After Jardine et al. (2005).

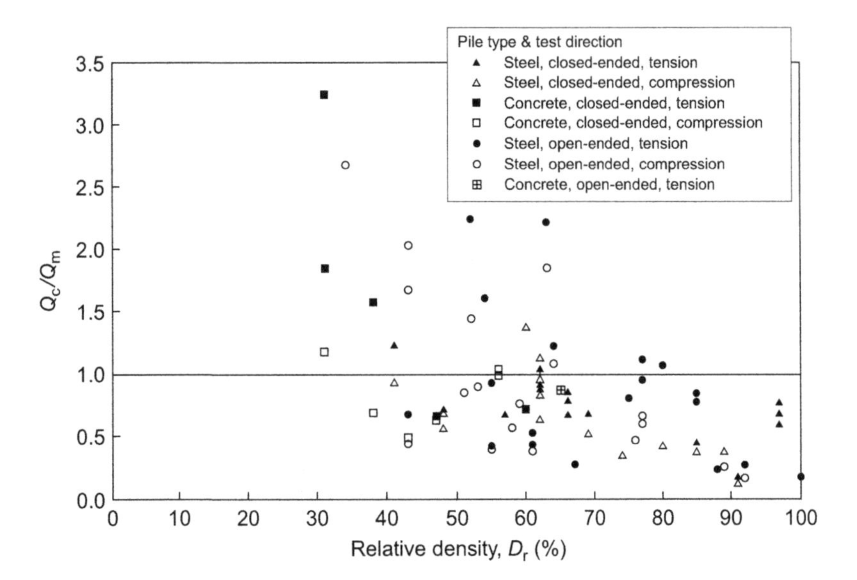

FIGURE 2.1 Distribution of Q_c/Q_m with respect to relative density, D_r: API (1993) shaft procedure for sands. *After Jardine et al. (2005).*

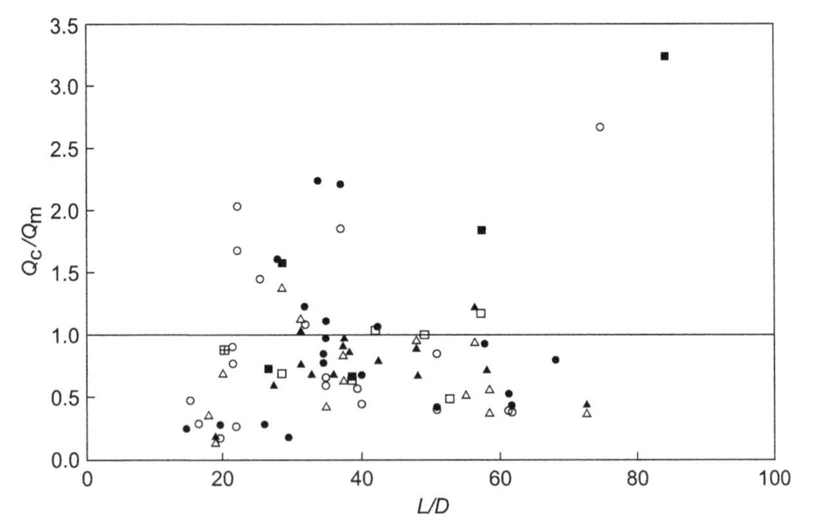

FIGURE 2.2 Distribution of Q_c/Q_m with respect to pile slenderness ratio, L/D: API (1993) shaft procedure for sands. *After Jardine et al. (2005).*

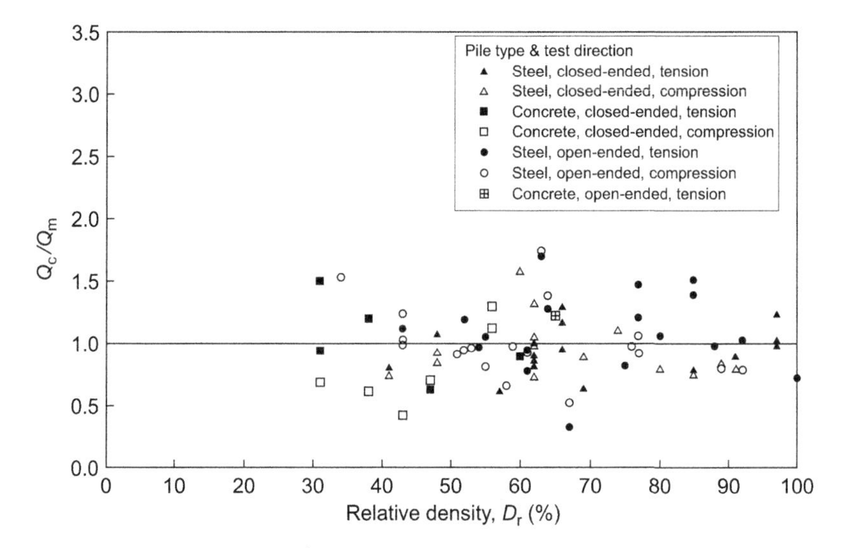

FIGURE 2.3 Distribution of Q_c/Q_m with respect to relative density, D_r: ICP shaft procedure for sands. *After Jardine et al. (2005).*

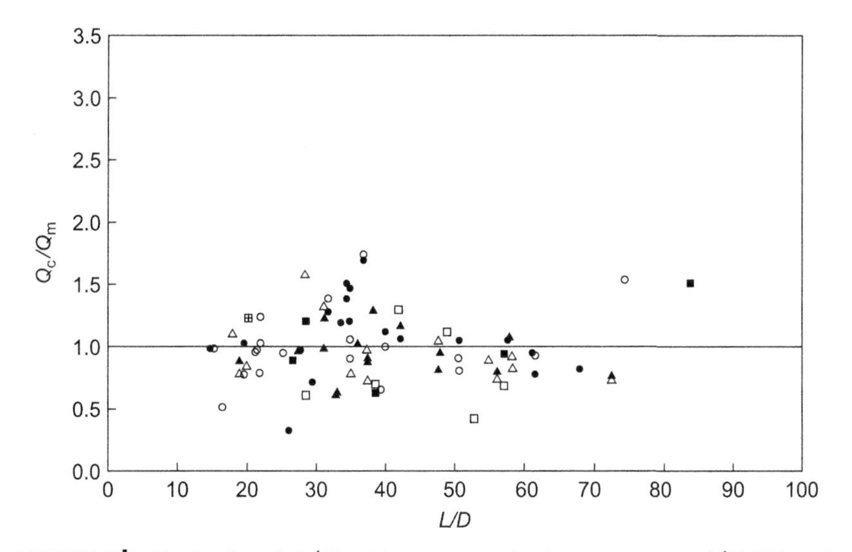

FIGURE 2.4 Distribution of Q_c/Q_m with respect to pile slenderness ratio, L/D: ICP shaft procedure for sands. *After Jardine et al. (2005).*

database. ICP-05 was equally applicable to open-ended and closed-ended piles and eliminated the strong skewing produced by the Main Text API method.

Lehane et al. (2005a) and Schneider et al. (2008) eliminated from the ICP database; some entries that did not meet their criteria and added further 26 tests, giving the "UWA" database summarized in Table 2.7 which was used to test the performances of the full versions of the "CPT-based" methods outlined above. The CPT methods offered considerable improvements over the API's Main Text

TABLE 2.9 Summary of UWA Database Assessment

Q_c/Q_m	API-00 (μ and ±CoV)	ICP-05 (μ and ±CoV)	UWA-05 (μ and ±CoV)	Fugro-05 (μ and ±CoV)	NGI-05 (μ and ±CoV)
All driven, open and closed	0.81 ± 0.67	0.95 ± 0.30	0.97 ± 0.27	1.11 ± 0.38	1.11 ± 0.37
Driven, open compression	0.75 ± 0.68	0.89 ± 0.28	0.98 ± 0.19	1.14 ± 0.30	1.01 ± 0.25
Driven, open tension	0.72 ± 0.76	0.90 ± 0.27	0.91 ± 0.23	0.90 ± 0.32	1.01 ± 0.35
Driven closed compression	0.78 ± 0.56	0.99 ± 0.33	0.98 ± 0.33	1.24 ± 0.39	1.16 ± 0.40
Driven closed tension	1.12 ± 0.84	1.02 ± 0.30	1.00 ± 0.29	0.97 ± 0.41	1.27 ± 0.50

After Lehane et al. (2005a).

method, as outlined in Table 2.9, considering only the mean μ and ±CoV values for Q_c/Q_m. Similar CoVs were noted for the "full" UWA and ICP approaches (0.19–0.33, depending on the cases considered) along with a generally marginally conservative bias, while NGI-05 appeared slightly nonconservative, especially for closed-ended piles and gave higher CoVs of 0.25–0.50. The Fugro-05 method gave outcomes between those found for the NGI and ICP/UWA methods. We are not aware of any systematic study being reported yet of how the "simplified" ICP or "offshore" UWA methods perform across comparably large databases of pile tests.

Significantly different statistical outcomes were reported by Kolk et al. (2005a) and Clausen et al. (2005) in their assessments of their respective methods. The outcomes appear to be sensitive to the pile test databases, especially for the larger piles. Taken together, the combined ICP and UWA databases comprised just over 100 different piles driven in silica sand and tested to failure. However, only 11 piles (from just three sites) were open-ended, had $D \geq 600$ mm and were accompanied by full CPT profiles. Further well-characterized pile tests (or perhaps enhanced CPT-based site investigations at sites where high-quality pile tests have been conducted recently) are required to augment this sparse data set, cover a wider range of international conditions and gain further insight into uncertain factors such as the effects of layering on base resistance (Xu, 2006) and the effects of pile age on capacity (Jardine et al., 2006; Karlsrud et al., 2014 or Gavin et al., 2013).

Chapter | Three

Description of the Extended ZJU-ICL Database

CHAPTER OUTLINE

3.1 OVERVIEW

The starting points for the new ZJU-ICL were the ICP and UWA databases. The ICP set reported by Jardine et al. (2005) added a significant number of new case histories to those assembled earlier by Lehane and Jardine (1994) and Chow (1997) to comprise 83 tests in sand. Schneider et al. (2008) augmented the ICP tests, adding 26 previously unrecognized entries. The UWA team also applied further quality filters, such as excluding any tests without full CPT profiles.

The new quality filters that we outline in Section 3.2 below led to 54 entries being adopted from the ICP database, along with 14 additional cases from the UWA set. The ZJU-ICL team has also assembled, to date, 48 further new test entries from their own projects (see Yang et al., 2015a), the literature, and through acknowledged communication with other research groups worldwide. The new cases contribute a 70% increase in the total population of tests that meet the current quality criteria. Tables 2.7 and 3.1–3.3 summarize the characteristics of the respective databases, while Table 3.4 provides a glossary for the terms employed.

Ideally, test piles should be instrumented to allow their shaft load distributions to be defined and the base capacities isolated in compression tests. A good spread of tension tests is also desirable. All of the ICP database entries adopted involved either strain gauged piles or tension tests. However, only 5 of the 14 new entries from the UWA database and 22 of the 48 new ZJU-ICL cases (including tension tests) allow shaft and base capacities to be separated.

TABLE 3.1 ICP Data Entries Adopted Into the ZJU-ICL Database

Test ID	Site name	Pile No.	Pile material	Pile shape	B or D (mm)	t (mm)	z_{tip} (m)	Water table depth (m)	Test type	Age (days)	Max Q_m (MN) [w] (mm)	0.1D Q_m (MN)	Clay contribution Q_{sc} (MN)	Average IFR (if available)	Interface friction angle δ_f (degrees)	Sources
001	Ogeechee River	H-2	C	S	406	–	15.2	1.5	C	0.5	3.16 [133]	2.75	–	–	Estimated by PSD	Vesic (1970)
002	Ogeechee River	H-12	S	C	457	–	6.1	1.5	C	0.5	2.14 [130]	2.08	–	–	Estimated by PSD	Vesic (1970)
003	Ogeechee River	H-13	S	C	457	–	8.9	1.5	C	0.5	2.81 [132]	2.64	–	–	Estimated by PSD	Vesic (1970)
004	Ogeechee River	H-14	S	C	457	–	12	1.5	C	0.5	3.56 [131]	3.21	–	–	Estimated by PSD	Vesic (1970)
005	Ogeechee River	H-15	S	C	457	–	15	1.5	C	0.5	4.12 [61]	3.95	–	–	Estimated by PSD	Vesic (1970)
006	Ogeechee River	H-16	S	C	609	–	15	1.5	T	1.5	1.54 [10]	1.54	–	–	Estimated by PSD	Vesic (1970)
007	Drammen	A	C	C	280	–	8	1.7	C	–	0.29 [39]	2.80	–	–	Default value	Gregersen et al. (1973)
008	Drammen	D/A	C	C	280	–	16	1.7	C	–	0.50 [41]	0.49	–	–	Default value	Gregersen et al. (1973)
009	Drammen	E-7.5	C	C	280	–	7.5	1.7	C	–	0.21 [32]	0.21	–	–	Default value	Gregersen et al. (1973)
010	Drammen	E-11.5	C	C	280	–	11.5	1.7	C	–	0.33 [30]	0.33	–	–	Default value	Gregersen et al. (1973)
011	Drammen	E-15.5	C	C	280	–	15.5	1.7	C	–	0.48 [35]	0.47	–	–	Default value	Gregersen et al. (1973)
012	Drammen	E-19.5	C	C	280	–	19.5	1.7	C	–	0.65 [32]	0.64	–	–	Default value	Gregersen et al. (1973)
013	Drammen	E-23.5	C	C	280	–	23.5	1.7	C	–	0.90 [40]	0.84	–	–	Default value	Gregersen et al. (1973)
014	Drammen	A(T)	C	C	280	–	8	1.7	T	–	0.09 [18]	0.09	–	–	Default value	Gregersen et al. (1973)
015	Drammen	D/A(T)	C	C	280	–	16	1.7	T	–	0.25 [37]	0.25	–	–	Default value	Gregersen et al. (1973)
016	Drammen	E-(T)	C	C	280	–	23.5	1.7	T	–	0.29 [37]	0.29	–	–	Default value	Gregersen et al. (1973)
017	Hoogzand	1-C	S	C	356	16	7	3.2	C	37	2.50 [64]	2.27	–	0.66	Estimated by PSD	Beringen et al. (1979)

018	Hoogzand	1-T	S	C	356	16	7	3.2	T	37	0.82 [20]	0.82	–	0.66	Estimated by PSD	Beringen et al. (1979)
019	Hoogzand	3-C	S	C	356	20	5.3	3.2	C	19	2.00 [64]	1.85	–	0.77	Estimated by PSD	Beringen et al. (1979)
020	Hoogzand	3-T	S	C	356	20	5.3	3.2	T	19	0.53 [10]	0.53	–	0.77	Estimated by PSD	Beringen et al. (1979)
021	Hoogzand	2-C	S	C	356	–	6.8	3.2	C	–	3.10 [64]	2.85	–	–	Estimated by PSD	Beringen et al. (1979)
022	Hoogzand	2-T	S	C	356	–	6.8	3.2	T	–	1.21 [57]	1.21	–	–	Estimated by PSD	Beringen et al. (1979)
023	Hunter's point	S	S	C	273	–	9.2	2.4	C	24	0.50 [83]	0.44	–	–	Default value	Briaud et al. (1989a)
024	Akasaka	6C	S	C	200	–	11	9	C	–	~1.5 [1000]	0.95	–	–	Estimated by PSD	BCP Committee (1971)
025	Hound point	P(0)-C	S	C	1220	24.2	26	0	C	21	7.50 [215]	7.00	0.44	0.95	Default value	Williams et al. (1997)
026	Hound point	P(0)-T1	S	C	1220	24.2	34	0	T	11	3.86 [25]	3.86	0.41	0.95	Default value	Williams et al. (1997)
027	Hound point	P(0)-T2	S	C	1220	24.2	41	0	T	4	3.74 [NA]	3.74	0.34	0.95	Default value	Williams et al. (1997)
028	Lemen	BD	S	C	660	19	38.1	0	T	–	5.25 [38.6]	5.25	–	0.84	Default value	Jardine et al. (1998)
029	Baghdad	P1-C	C	S	253	–	11	6.2	C	88	1.10 [110]	0.95	–	–	Estimated by PSD	Altaee et al. (1992)
030	Baghdad	P1-T	C	S	253	–	11	6.2	T	200	0.58 [65]	0.58	–	–	Estimated by PSD	Altaee et al. (1992)
031	Baghdad	P2-C	C	S	253	–	15	6	C	42	1.61 [29]	1.61	–	–	Estimated by PSD	Altaee et al. (1992)
032	Dunkirk	CL-T	S	C	324	12.7	11.3	4	T	175	0.44 [50]	0.44	–	0.72	From ring shear test	Chow (1997)
033	Dunkirk	CS-T	S	C	324	12.7	11.3	4	T	187	0.40 [60]	0.40	–	0.72	From ring shear test	Chow (1997)
034	Dunkirk	R1-T	S	C	457	13.5	19.3	4	T	9	1.45 [31]	1.45	–	0.78	From ring shear test	Jardine et al. (2006)
035	Dunkirk	C1-C	S	C	457	13.5	10	4	C	68	2.82 [34]	2.82	–	0.78	From ring shear test	Jardine et al. (2006)
036	Dunkirk	C1-T	S	C	457	13.5	10	4	T	69	0.82 [46]	0.82	–	0.78	From ring shear test	Jardine et al. (2006)
037	Euripides	Ia	S	C	763	35.6	30.5	1	C	7	11.6 [260]	7.40	–	0.99	From ring shear test	Kolk et al. (2005b)

Continued...

TABLE 3.1 ICP Data Entries Adopted Into the ZJU-ICL Database—continued

Test ID	Site name	Pile No.	Pile material	Pile shape	B or D (mm)	t (mm)	z_{tip} (m)	Water table depth (m)	Test type	Age (days)	Max Q_m (MN) [w] (mm)	$0.1D$ Q_m (MN)	Clay contribution Q_{sc} (MN)	Average IFR (if available)	Interface friction angle δ_f (degrees)	Sources
038	Euripides	Ib	S	C	763	35.6	38.7	1	C	2	16.26 [249]	13.00	–	0.97	From ring shear test	Kolk et al. (2005b)
039	Euripides	Ic	S	C	763	35.6	47	1	C	11	23.41 [260]	18.80	–	0.96	From ring shear test	Kolk et al. (2005b)
040	Euripides	Ia-T	S	C	763	35.6	30.5	1	T	7	1.66 [76]	1.66	–	0.99	From ring shear test	Kolk et al. (2005b)
041	Euripides	Ib-T	S	C	763	35.6	38.7	1	T	2	8.40 [36]	8.40	–	0.97	From ring shear test	Kolk et al. (2005b)
042	Euripides	Ic-T	S	C	763	35.6	47	1	T	11	12.50 [72]	12.50	–	0.96	From ring shear test	Kolk et al. (2005b)
043	Euripides	II	S	C	763	35.6	46.7	1	C	6	21.53 [190]	18.40	–	0.95	From ring shear test	Kolk et al. (2005b)
044	Euripides	II-T	S	C	763	35.6	46.7	1	T	7	9.50 [76]	9.50	–	0.95	From ring shear test	Kolk et al. (2005b)
045	Lock and Dam 26	3-1	S	C	305	–	14.2	0	C	35	1.32 [76]	1.17	–	–	Estimated by PSD	Briaud et al. (1989b)
046	Lock and Dam 26	3-4	S	C	356	–	14.4	0	C	27	1.13 [33]	1.15	–	–	Estimated by PSD	Briaud et al. (1989b)
047	Lock and Dam 26	3-7	S	C	406	–	14.6	0	C	28	1.79 [76]	1.62	–	–	Estimated by PSD	Briaud et al. (1989b)
048	Lock and Dam 26	3-2	S	C	305	–	11	0	T	35	0.54 [62]	0.54	–	–	Estimated by PSD	Briaud et al. (1989b)
049	Lock and Dam 26	3-5	S	C	356	–	11.1	0	T	27	0.61 [43]	0.61	–	–	Estimated by PSD	Briaud et al. (1989b)
050	Lock and Dam 26	3-8	S	C	406	–	11.1	0	T	28	0.90 [60]	0.90	–	–	Estimated by PSD	Briaud et al. (1989b)
051	Tokyo-bay	TP	S	C	2000	34	30.6	0	C	52	34.68 [203]	34.68	1.32	1.00	Default value	Shioi et al. (1992)
052	Hsin Ta	TP 4	S	C	609	–	34.3	2	C	33	4.26 [78]	4.26	0.72	–	Default value	Yen et al. (1989)
053	Hsin Ta	TP 5	S	C	609	–	34.3	2	T	28	2.45 [21]	2.63	0.18	–	Default value	Yen et al. (1989)
054	Hsin Ta	TP 6	S	C	609	–	34.3	2	C	30	4.40 [21]	4.91	0.61	–	Default value	Yen et al. (1989)

TABLE 3.2 UWA Data Entries Adopted Into the ZJU-ICL Database

Test ID	Site name	Pile No.	Pile material	Pile shape	B or D (mm)	t (mm)	z_{tip} (m)	Water table depth (m)	Test type	Age (days)	Max Q_m (MN) [w] (mm)	0.1 D Q_m (MN)	Clay contribution Q_{sc} (MN)	Average IFR (if available)	Interface friction angle δ_f	Sources
001	Drammen	16-P1-11	S	C	813	12.5	11	3	C	2	1.60 [204]	1.21	–	0.88	Default value	Tveldt and Fredriksen (2003)
002	Drammen	25-P2-15	S	C	813	12.5	15	3	C	2	2.05 [NA]	1.89	–	0.88	Default value	Tveldt and Fredriksen (2003)
003	Drammen	25-P2-25	S	C	813	12.5	25	3	C	2	3.28 [NA]	2.70	–	0.88	Default value	Tveldt and Fredriksen (2003)
004	Shanghai	ST-1	S	C	914	20	79	0.5	C	23	16.36 [121]	15.56	1.72	0.80	Default value	Pump et al. (1998)
005	Shanghai	ST-2	S	C	914	20	79.1	0.5	C	35	17.82 [130]	17.08	1.74	0.85	Default value	Pump et al. (1998)
006	Cimarron River	p1	S	C	660	–	19	1	C	–	3.58 [80]	3.57	–	–	Default value	Nevels and Snethen (1994)
007	Cimarron River	p2	C	0	610	–	19.5	1	C	–	3.56 [65]	3.56	–	–	Default value	Nevels and Snethen (1994)
008	Jonkoping	P23	C	S	235	–	16.8	1.3	C	>1	1.72 [50]	1.72	0.06	–	Default value	Jendeby et al. (1994)
009	Jonkoping	P25	C	S	235	–	17.8	1.3	C	<1	1.50 [14]	1.65	0.06	–	Default value	Jendeby et al. (1994)
010	Jonkoping	P26	C	S	275	–	16.2	1.3	C	>1	1.40 [50]	1.36	0.07	–	Default value	Jendeby et al. (1994)
011	Fittja Straits	D-5	C	S	235	–	12.8	2	C	5	0.36 [35]	0.34	0.03	–	Estimated by PSD	Axelsson (2000)
012	Sermide	S	S	C	508	–	35.9	0	C	–	5.62 [84]	5.49	0.29	–	Default value	Appendino (1981)
013	Pigeon Creek	1	S	C	356	–	6.9	3	C	4	1.77 [150]	1.50	0.01	–	Default value	Paik et al. (2003)
014	Pigeon Creek	2	S	C	356	32	7	3	C	4	1.28 [135]	1.03	0.02	0.83	Default value	Paik et al. (2003)

TABLE 3.3 New Entries Added to the ZJU-ICL Database

Test ID	Site name	Pile No.	Pile material	Pile shape	*B* or *D* (mm)	*t* (mm)	z_{tip} (m)	Water table depth (m)	Test type	Age (days)	Max Q_m (MN) [*w*] (mm)	0.1*D* Q_m (MN)	Clay contribution Q_{sc} (MN)	Average IFR (if available)	Interface friction angle δ_f	Sources
001	Wuhu	K24-1	C	C	600	130	33	0.72	C	5	2.40 [86]	2.00	0.03	0.74	Estimated by PSD	Yang et al. (2015a)
002	Wuhu	K24-2	C	C	600	130	39.8	0.72	C	14	4.80 [86]	4.40	0.23	0.74	Estimated by PSD	Yang et al. (2015a)
003	Wuhu	K24-3	C	S	500	127	39.8	0.72	C	13	4.80 [84.9]	4.45	0.23	0.73	Estimated by PSD	Yang et al. (2015a)
004	Wuhu	K34-1	C	C	600	130	29.3	1.1	C	15	5.40 [85]	4.90	0.07	0.82	Estimated by PSD	Yang et al. (2015a)
005	Wuhu	K27-1	C	C	800	130	29.2	0.3	C	13	5.40 [87]	5.27	0.04	0.74	Estimated by PSD	Yang et al. (2015a)
006	Rio de Janeiro	PI-1	C	C	500	–	37.2	2.7	C	64	3.59 [51]	3.59	0.05	–	Default value	Tsuha (2012)
007	Rio de Janeiro	PI-2a	C	C	500	–	21.4	2.7	C	72	1.95 [32]	1.95	0.05	–	Default value	Tsuha (2012)
008	Rio de Janeiro	PI-3	C	C	700	–	35.6	2.53	C	89	6.01 [21]	–	0.05	–	Default value	Tsuha (2012)
009	Rio de Janeiro	PI-4	C	C	500	–	26.5	2.42	C	86	4.55 [82]	4.53	0.04	–	Default value	Tsuha (2012)
010	Dublin	S2	S	C	340	14	7	13	T	2	0.34 [34]	0.34	–	0.73	From ring shear test	Gavin et al. (2013)
011	Dublin	S3	S	C	340	14	7	13	T	13	0.67 [59]	0.67	–	0.73	From ring shear test	Gavin et al. (2013)
012	Dublin	S5	S	C	340	14	7	13	T	220	0.99 [20]	0.99	–	0.73	From ring shear test	Gavin et al. (2013)
013	Horstwalde	P2B	S	C	711	12.5	17.61	0	T	43	1.40 [17]	1.40	–	0.86	Default value	Rücker et al. (2013)
014	Horstwalde	P2D	S	C	711	25	17.69	0	T	34	1.40 [19]	1.40	–	0.85	Default value	Rücker et al. (2013)
015	Horstwalde	P5B	S	C	711	12.5	17.71	0	T	36	1.42 [10]	1.42	–	0.86	Default value	Rücker et al. (2013)
016	Horstwalde	P5D	S	C	711	12.5	17.76	0	T	29	0.95 [18]	0.95	–	0.86	Default value	Rücker et al. (2013)
017	Horstwalde	P4B	S	C	711	12.5	17.67	0	T	37	1.55 [10]	1.55	–	0.86	Default value	Rücker et al. (2013)

018	Horstwalde	P4D	S	C	711	12.5	17.66	0	T	32	1.25 [37]	1.25	–	0.86	Default value	Rücker et al. (2013)
019	Horstwalde	P3B	S	C	711	12.5	17.63	0	T	116	1.90 [10]	1.90	–	0.86	Default value	Rücker et al. (2013)
020	Horstwalde	P3D	S	C	711	12.5	17.74	0	T	30	1.12 [5]	1.12	–	0.86	Default value	Rücker et al. (2013)
021	Columbia	P1	S	C	610	–	45	0	C	15	4.00 [134]	3.75	0.06	–	Default value	Naesgaard et al. (2012)
022	Hampton River	P1	C	S	610	–	16.8	0	C	12	3.1 [20]	3.10	–	–	Default value	Pando et al. (2003)
023	Rotterdam	P6	C	S	380	–	30.6	0	C	–	4.32 [41]	4.32	–	–	Default value	de Gijt et al. (1995)
024	Rotterdam	P8	C	S	380	–	30.3	0	C	–	4.65 [50]	4.45	–	–	Default value	de Gijt et al. (1995)
025	Rotterdam	P10	C	S	380	–	30.7	0	C	–	4.33 [39]	4.33	–	–	Default value	de Gijt et al. (1995)
026	Waddinxveen	P2	C	S	350	–	10	10	C	33	1.15 [61]	1.15	0.01	–	Default value	Hölscher (2009)
027	Mobile Bay	AL 1	S	C	324	25.4	15.2	0	C	–	1.25 [24]	1.25	–	0.71	Default value	Mayne (2013)
028	Mobile Bay	AL 2	S	C	324	25.4	42.7	0	C	–	3.75 [96]	3.35	–	0.71	Default value	Mayne (2013)
029	ABEF Foundation	7	C	C	500	90	9.0	42.6	C	–	3.2 [66]	3.14	–	0.73	Default value	Mayne (2013)
030	ABEF Foundation	8	C	C	500	90	7.5	14.7	C	–	3.3 [125]	3.14	–	0.73	Default value	Mayne (2013)
031	Apalachicola	BR 1	C	S	610	–	29.9	0	C	–	4.31 [32]	4.31	–	–	Default value	Mayne (2013)
032	Los Angeles	CA	C	S	610	–	29	0	C	–	5.69 [70]	5.60	–	–	Default value	Mayne (2013)
033	MS Smith	1045	C	S	410	–	10.2	0	C	–	1.96 [10]	1.96	–	–	Default value	Mayne (2013)
034	MS Desota	2108	C	S	460	–	7.6	0	C	–	1.60 [48]	1.60	–	–	Default value	Mayne (2013)
035	MS Harrison	3028	C	S	460	–	16.2	0	C	–	1.43 [28]	1.43	–	–	Default value	Mayne (2013)
036	Washington	3118A	C	S	410	–	7.6	0	C	–	0.75 [27]	0.75	–	–	Default value	Mayne (2013)

Continued...

TABLE 3.3 New Entries Added to the ZJU-ICL Database—continued

Test ID	Site name	Pile No.	Pile material	Pile shape	B or D (mm)	t (mm)	z_{tip} (m)	Water table depth (m)	Test type	Age (days)	Max Q_m (MN) [w] (mm)	$0.1D$ Q_m (MN)	Clay contribution Q_{sc} (MN)	Average IFR (if available)	Interface friction angle δ_f	Sources
037	Washington	3123B	C	S	360	–	16.6	0	C	–	1.10 [28]	1.10	–	–	Default value	Mayne (2013)
038	Washington	3142A	C	S	360	–	6.2	0	C	–	0.92 [53]	0.85	–	–	Default value	Mayne (2013)
039	Larvik site	L1	S	C	508	6.3	21.5	2	T	43	0.98 [18]	0.98	–	0.80	Default value	Karlsrud et al. (2014)
040	Larvik site	L2	S	C	508	6.3	21.5	2	T	135	0.99 [37]	0.98	–	0.80	Default value	Karlsrud et al. (2014)
041	Larvik site	L3	S	C	508	6.3	21.5	2	T	218	1.16 [31]	0.98	–	0.80	Default value	Karlsrud et al. (2014)
042	Larvik site	L4	S	C	508	6.3	21.5	2	T	365	1.07 [48]	0.98	–	0.80	Default value	Karlsrud et al. (2014)
043	Larvik site	L5	S	C	508	6.3	21.5	2	T	730	1.08 [50]	0.98	–	0.80	Default value	Karlsrud et al. (2014)
044	Larvik site	L6	S	C	508	6.3	21.5	2	T	730	0.9 [32]	0.98	–	0.80	Default value	Karlsrud et al. (2014)
045	Larvik site	L7	S	C	508	6.3	21.5	2	T	30	0.6 [38]	0.6	–	0.80	Default value	Karlsrud et al. (2014)
046	Jackson County	JCEPF 2	S	C	273	–	17.8	16	C	10	1.09 [8]	1.09	–	–	Default value	Mayne and Elhakim (2002)
047	Lafayette BRG	MAT2	S	C	356	–	20.3	5	C	54	2.23 [20]	2.23	–	–	Default value	Komurka and Grauvogl-Graham (2010)
048	Fittja Straits	D-1	C	S	235	–	13.0	2	C	1	0.31 [31]	0.31	0.03	–	Estimated by PSD	Axelsson (2000)

TABLE 3.4 Glossary of Terms Used in Tables 3.1–3.3

Column	Description
Test ID	ID number for case in ZJU-ICL database
Site	Site name shown in source
Pile No.	Pile ID number
Pile material	Material from which pile was made; C = concrete and S = steel
Pile shape	Exterior shape of pile; C = circular; S = square; O = octagonal
B or D	Outer width of square pile or octagonal piles or diameter of circular piles
t	Wall thickness for open-ended pile
z_{tip}	Tip depth of pile
Water table depth	Depth to water table at time of driving
Test type	Compression or tension test; C = compression and T = tension
Age	Time of load testing after pile was driven
Max Q_m	Maximum load measured in pile load test
$[w]$	Maximum displacement measured in pile load test
0.1 D Q_m	Measured pile capacity at $w = 0.1\,D$
Clay contribution Q_{sc}	Predicted clay shaft capacity contribution
IFR	Incremental filling ratio of open-ended pile
δ_f	Interface friction angle with default value = 29°
Source	The source of the pile load test information

Each case is entered in the database following a format similar to that adopted by Niazi (2014). An example entry from the authors' research at the Wuhu Yangtze River Bridge site in China (Yang et al., 2015a) is given in Figure 3.1. The complete data entries are given in the appendix.

Figure 3.2 illustrates the geographical distribution of the new combined data set, which increases the number of countries considered from 10 to 13. The ZJU-ICL team will update the database periodically adding any submitted entries that meet the above criteria and data quality levels provided in the appendix. All such entries will be acknowledged fully. Expanding the database will increase its value as an inclusive and freely available international research resource.

3.2 QUALITY CRITERIA

The ZJU-ICL database applies similar, but marginally more stringent, criteria to the earlier ICP and UWA studies. All entries must:

1. comprise tests where more than 65% of the shaft capacity and all of any end bearing is developed within silica sand strata.

(a)

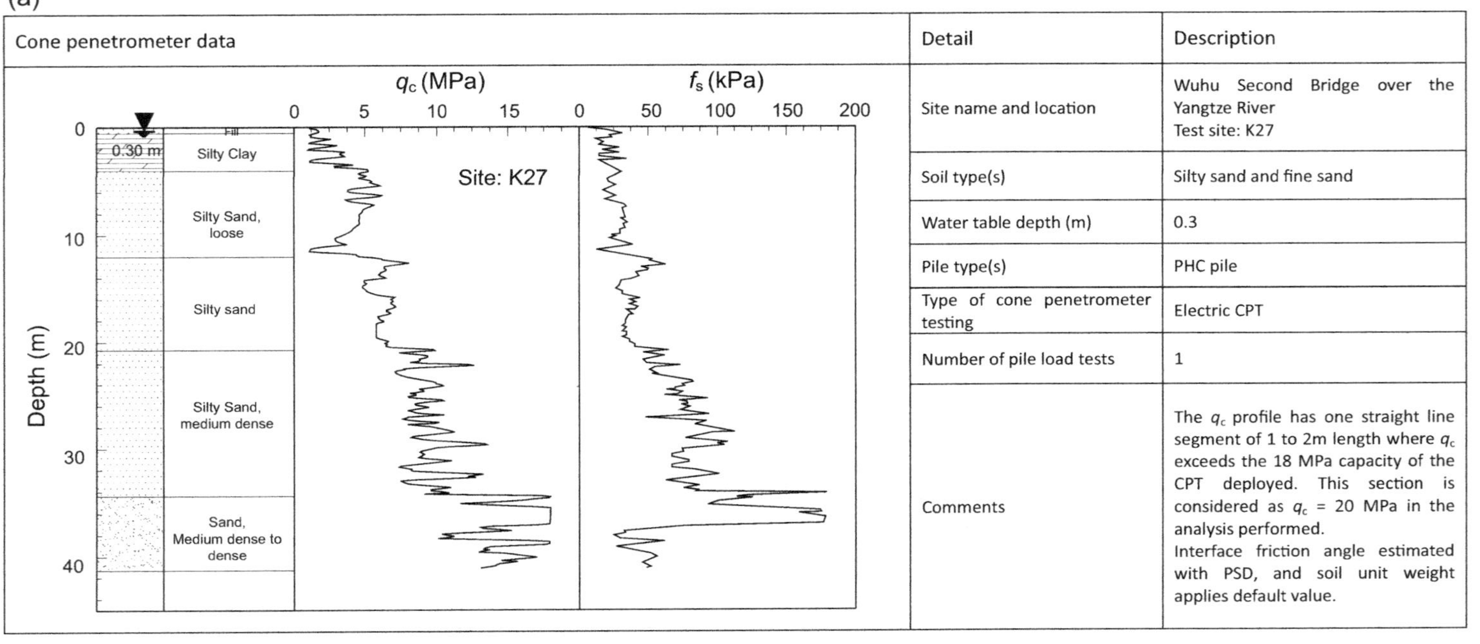

Detail	Description
Site name and location	Wuhu Second Bridge over the Yangtze River Test site: K27
Soil type(s)	Silty sand and fine sand
Water table depth (m)	0.3
Pile type(s)	PHC pile
Type of cone penetrometer testing	Electric CPT
Number of pile load tests	1
Comments	The q_c profile has one straight line segment of 1 to 2m length where q_c exceeds the 18 MPa capacity of the CPT deployed. This section is considered as q_c = 20 MPa in the analysis performed. Interface friction angle estimated with PSD, and soil unit weight applies default value.

(b)

Load–displacement data

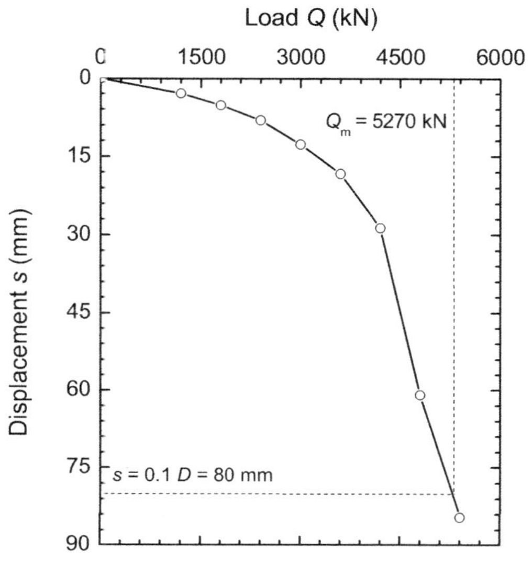

Detail	Description
Pile type/material	Open-ended concrete pile
Length, L (m)	29.2
Outer diameter, D/B (mm)	800 (130)
Installation method	Driven
Set up time, days	13
Loading mode	Compression
$Q_{max\text{-}measured}$ (kN)	5400
$Q_{0.1D}$ (kN)	5270
Q_s (kN)	Not isolated
Q_b (kN)	Not isolated
API Q_c (kN)	3607
Q_c/Q_m	0.68
UWA Q_c (kN)	4536
Q_c/Q_m	0.86
ICP Q_c (kN)	5053
Q_c/Q_m	0.96
Fugro Q_c (kN)	6006
Q_c/Q_m	1.14
NGI Q_c (kN)	4282
Q_c/Q_m	0.81

FIGURE 3.1 An example of description of test site and pile load tests (a) Site ID No. 3: K27, Wuhu, China. Yang et al. (2015a). (b) Pile ID: Wuhu K27-1.

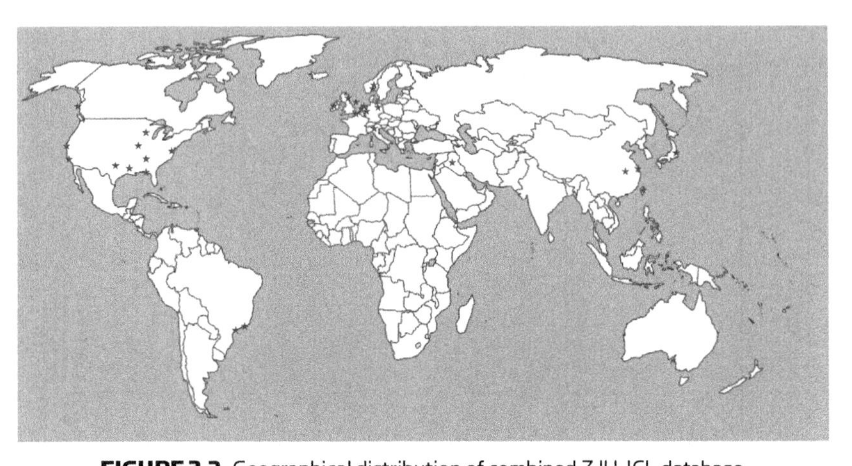

FIGURE 3.2 Geographical distribution of combined ZJU-ICL database.

2. be accompanied by an adequate site investigation including a complete CPT profile from a nearby location, soil descriptions, information on groundwater levels and sand particle size distribution. Ideally, good measurements of in situ density and interface shearing angles should also be available. Only silica sand sites may be included, i.e., carbonate sand sites are ignored.
3. give records from high-quality (ideally load-controlled) first-time axial tests to failure, including load–displacement curves that continue until either peak loads or axial displacements of 0.1 D have developed. Retests on the same piles are generally not included, the exceptions being cases where piles were tested shortly after driving to more than one depth, where the depth intervals are sufficiently different to reduce possible interactions between successive tests.[1] Tests equipped with strain gauges and tension tests to failure are particularly valuable for isolating the shaft-to-base capacity splits from compression tests.
4. offer information on driving method, pile embedment, diameter, tip end conditions, wall thickness, and material. Ideally, the pile driving records and pile age after driving should also be available. The database is then divided into a main set with pile ages of 10–100 days and a subset of tests conducted at both earlier and later ages. Tests for which the age on testing is uncertain are assumed to fall in the usual 10–100 day range.

Axial capacity is known to be significantly influenced by the time elapsed between pile installation and testing; Jardine et al. (2006), Gavin et al. (2013), Karlsrud et al. (2014), Rimoy et al. (2015). Figure 3.3 presents the Q_m/Q_c ratios evaluated from ICP-05 calculations, plotted against time after installation covering the 3–300 day age range. As shown later in Section 5.4, base capacity is insensitive to time, leading shaft capacity to be more affected than the total.

[1] A case may also be made for including tension tests conducted shortly after compression tests on freshly driven piles, as the compression test might be considered analogous to a final blow. However, any such correspondence cannot be assumed to apply to piles tested after any aging interval.

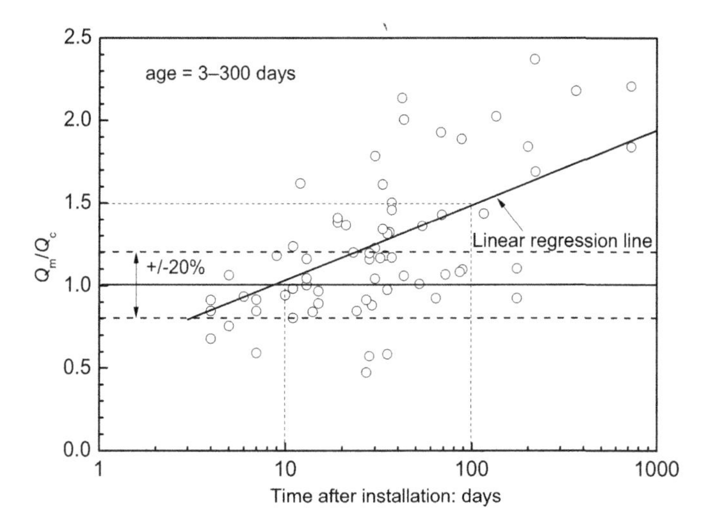

FIGURE 3.3 Q_m/Q_c from ICP-05 against time after installation.

A linear regression suggests that the ICP calculations provide a best fit at around 10–15 days, with capacity growing by approximately 50% per log cycle of time over the next 100–200 days. This trend introduces significant statistical bias that could be reduced by introducing tight tolerance limits to the test ages. If, for example, an age range of just 3–23 days was considered, the effects of time could be kept within ±20%. However, such a step would reduce the number of pile tests drastically. Balancing the requirements of maintaining a sufficient number of tests with those of to keeping age effects within tolerable limits, the ZJU-ICL database has opted to apply a 10–100 day age range in its filtering of entries into subsets that lie both within and without the specified age limits.

3.3 DETAILED CHARACTERISTICS

Table 3.5 summarizes the general characteristics of the ZJU-ICL database. Applying the 10–100 day age filter leads to a total of 35 tests that fall outside the limits. Of the 80 remaining tests, 48 involved closed-end piles while 32 had open-ends. The tests were conducted in a total of 13 countries and 4 continents; 44 cases in this subset, including the 24 tension tests, allow shaft and base capacities to be separated.

Figure 3.4 presents histograms that illustrate the distributions of pile age on testing, total capacity, nominal pile diameter/width, pile length, and average relative density along shaft and at toe. These plots suggest that:

1. the known pile test ages fall principally in the 10–100 day range. We assume that this range applies to the 35 (out of 116) entries for which the test age is unknown.
2. most piles developed capacities below 6 MN, with only 6 tests exceeding 10 MN.

TABLE 3.5 Summary of ZJU-ICL Database

	All entries			Filtered entries with age = 10–100 days		
	Closed	Open	All	Closed	Open	All
Number of piles	61	55	116	48	32	80
Steel	24	48	72	18	26	44
Concrete	37	7	44	30	6	36
Tension tests	10	31	41	8	16	24
Compression tests	51	24	75	40	16	56
Average length L (m)	17.6	25.2	21.2	18.9	26.0	21.8
Range of lengths L (m)	6.2–45	5.3–79.1	5.3–79.1	6.2–45	5.3–79.1	5.3–79.1
Average of diameter D (m)	0.413	0.645	0.522	0.422	0.667	0.520
Range of diameter D (m)	0.2–0.7	0.324–2.0	0.2–2.0	0.2–0.7	0.324–2.0	0.2–2.0
Average of density D_r (%)	54	60	57	54	61	57
Range of D_r (%)	28–89	30–88	28–89	31–89	30–87	30–89
Average test time after installation (days)	35	80	61	43	28	35

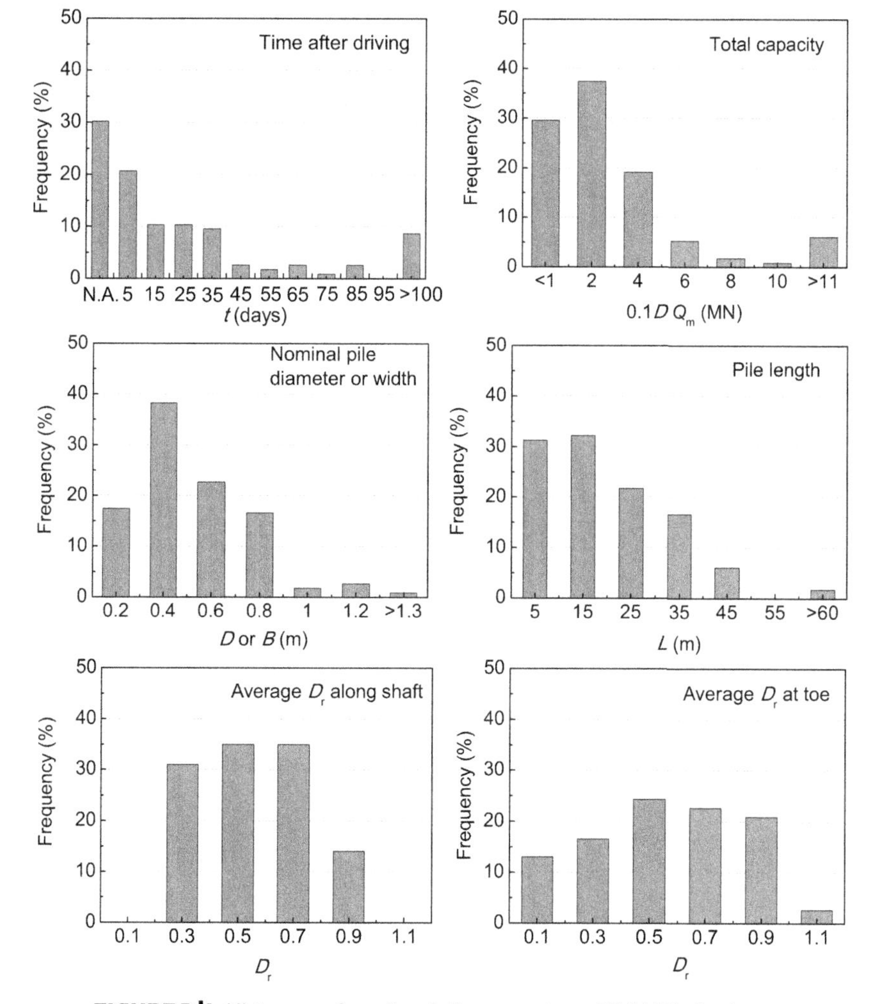

FIGURE 3.4 Histograms for soil and pile parameters of ZJU-ICL database.

3. the most common diameter and length ranges are 400–800 mm and 10–45 m, respectively.
4. the average relative densities estimated along the shafts and at the toes classify as medium to very dense in most cases, although the pile tips appeared to be positioned in loose layers with $D_r < 20\%$ in 13% of cases. In total 32 tests (at 7 sites) involved sands whose relatively loose states fall outside the range over which the API Main Text approach is currently recommended as being applicable.

Chapter | Four

Calculation Methods Applied in Database Analysis

CHAPTER OUTLINE

4.1 APPROACH

The main purpose of this booklet is to introduce and explain the origins and characteristics of the ZJU-ICL pile database. However, the information gathered has also been employed in a preliminary analysis of how predictions made with the API Main Text and six "2005 CPT-based methods" (comprising ICP and its simplified version, UWA and its offshore version as well as the Fugro and NGI methods) compare with the field test data. The calculation procedures applied are outlined below before detailing how parameter assessment was made for each of the methods. The cases adopted into the ZJU-ICL database from the ICP and UWA databases have been recalculated following the same procedures to ensure consistency, adopting where possible more refined CPT q_c values and calculation resolution. This step also provided a means of checking the results obtained and eliminating any errors. Chapter 5 presents a preliminary statistical analysis of the results obtained.

A Comprehensive Database of Tests on Axially Loaded Piles Driven in Sand

4.2 PROCEDURES

The Microsoft Excel pile capacity calculations involved 10 main steps.

1. Enter the basic information, such as site location, groundwater table, pile length, diameter, type, loading type, measured pile capacity and load–displacement curve.
2. Extract interpreted CPT q_c data at intervals of 0.1–0.2 m intervals, depending on pile length.
3. Interpret the soil layering based on site investigation borehole logs and CPT profile.
4. Assign unit weight for each layer. A default value of $\gamma = 19\,kN/m^3$ applied if the unit weight was uncertain.
5. Assign the interface shearing angles δ_f for ICP-05 and UWA-05. For the API method, β and limiting τ values are applied based on Table 2.2. For loose sand cases, the K_f and δ_f angles and τ limits specified in API (2000) are applied.
6. Interpolate CPT q_c and γ over 0.1 m or 0.2 m depth intervals.
7. Account for pore water pressures for soil layers beneath the groundwater table and calculate the free field effective σ'_{v0} at each depth.
8. Estimate nominal relative density D_r based on CPT q_c value, as required in API-00, ICP-05, and NGI-05 for varying purposes. In API-2014, D_r is used to determine the state of the sand and select the appropriate parameters for calculations. In ICP-05, D_r is employed to estimate the pile end condition, plugged, or unplugged. In NGI-05, D_r is used to calculate the parameter F_{D_r} in Table 2.6.
9. Average the q_c values used in CPT methods. For ICP-05 and Fugro-05, $q_{c,avg}$ is the average value of q_c over $\pm 1.5D$ at tip level. UWA-05 adopts Dutch averaging technique while NGI-05 uses the q_c recorded at the tip to find D_r as specified by NGI.
10. Calculate the shaft (Q_s), base (Q_b), and total (Q_{total}) capacities of piles for these seven methods. Following Schneider et al. (2008), the unit shaft resistances in any thin clay layers were assumed equal to be $q_t/35$ for simplicity and consistency, in which q_t is CPT end resistance with pore pressure correction for piezocones.

4.3 KEY PARAMETERS AND THEIR ASSESSMENT

The selection of key parameters has been carried out as described below:

4.3.1 Relative Density D_r

Relative density D_r is employed in the API-2014, ICP-05, and NGI-05 calculations for various purposes. The following empirical relationship proposed by Jamiolkowski et al. (2003) is used to estimate the relative density D_r from CPT q_c values for the API-00 and ICP-05 calculations:

$$D_r = 0.35 \ln(q_{c1N}/20) \tag{4.1a}$$

where q_{c1N} is the normalized cone tip resistance and can be related to q_c via the following equation,

$$q_{c1N} = (q_c/p_a)/(\sigma'_{v0}/p_a)^{0.5},\ p_a = 100\ \text{kPa} \quad (4.1b)$$

A slightly different equation is used in NGI-05,

$$D_r = 0.4\ln(q_{c1N}/22) \quad (4.2)$$

4.3.2 Incremental Filling Ratio and Final Filling Ratio

Both incremental filling ratio (IFR) and final filling ratio (FFR) pile coring parameters are used in the full UWA-05 procedures for open-ended piles. IFR is defined by,

$$\text{IFR} = \Delta L_p/\Delta z \quad (4.3)$$

where ΔL_p is the increment of soil plug length, and Δz is the increment of the pile penetration. Ideally, IFR should be recorded during driving. If it is not measured, the following equation can be used to estimate IFR:

$$\text{IFR} = \min[1, (D_i/1.5)^{0.2}] \quad (4.4)$$

where D_i has the unit of meter (m). FFR is the average value of IFR recorded over the last 3D of pile penetration.

4.3.3 Interface Shearing Angle δ_f

The constant volume (ultimate) interface shearing angle δ_f used in ICP-05 and UWA-05 or their simplified versions is a key parameter that can vary with sand grain size, shape, and mineral type as well as the hardness and roughness of the pile surface. Site-specific ring shear tests provide the most reliable data. However, Jardine and Chow (1996) originally recommended that if such tests are unavailable, trends for δ_f with grain d_{50} could be used to provide estimates, and reported curves from constant normal stress direct shear tests on a range of clean standard sands shearing against steel interfaces prepared with initial R_a values between 6 and 10 μm, after Jardine et al. (1992). The latter trends showed δ_f falling with d_{50} as shown in Figure 4.1. This recommendation was revised by Jardine et al. (2005) who reported flatter trends with d_{50} as shown in Figure 4.1 based on ring shear interface tests and also direct shear tests on offshore sands conducted for Shell (UK).

Noting that the pile driving in dense sand abrades the surface of steel piles and greatly reduce their maximum roughness; CUR (2001) proposed to adopt a constant $\delta_f = 29°$, irrespective of grain size. This recommendation is incorporated into the Fugro-05 procedures and provides a default for the UWA method. More recently, Ho et al. (2011) presented a systematic study of large-displacement ring shear interface tests on a wide range of materials

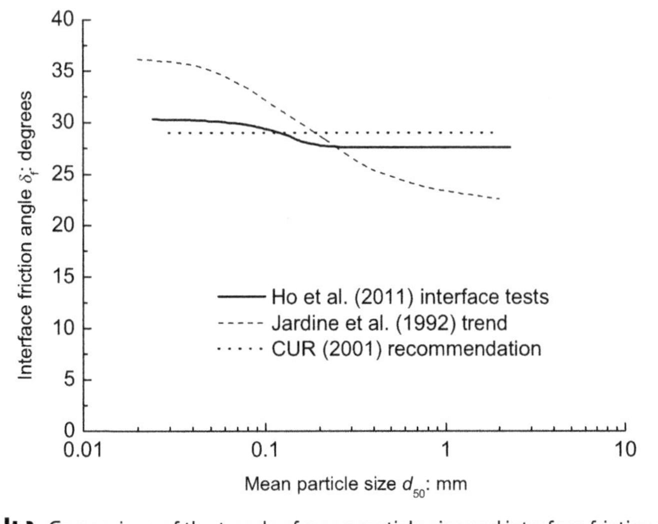

FIGURE 4.1 Comparison of the trends of mean particle size and interface friction angle for silica sand.

with d_{50} varying from 0.04 to 1.46 mm. These tests, which simulated field shaft abrasion and grain breakage, showed that the large-displacement interface angles δ_f reduces with d_{50}, but far less markedly than in Jardine et al.'s (1992) direct shear tests; see Figure 4.1. The ZJU-ICL database adopts this new curve for any ICP or UWA calculations for steel piles for which no other better data are available. The Fugro calculations retain the specified 29° angle. Concrete interfaces do not necessarily develop the same interface angles as those made of steel. Ring shear experiments reported by Barmpopoulos et al. (2009) involving clean silica sands and concrete interfaces indicate large-displacement interface shear resistance angles that followed slightly different pile roughness-to-soil d_{50} ratios, as shown in the summary diagram replotted as Figure 4.2. The suggested trend supports an assumption for rough piles of $\delta_f = 29°$ which coincidentally matches the default value proposed for steel piles in the Fugro-05 method.

However, site-specific ring shear interface tests remain the recommended ICP approach for establishing δ_f values for any given practical application.

4.3.4 Soil Unit Weight γ and Water Table

The unit weight of soil γ is used to calculate the overburden effective stress. While values for γ can be often found in site investigation reports, they may be affected by sampling disturbance. Checks are therefore made against the relative densities expected from the CPT profiles, and a default value of 19 kN/m^3 is adopted if credible site-specific data are not available. It is also imperative that the groundwater table is known from piezometers, standpipes, piezocone, or drilling measurements.

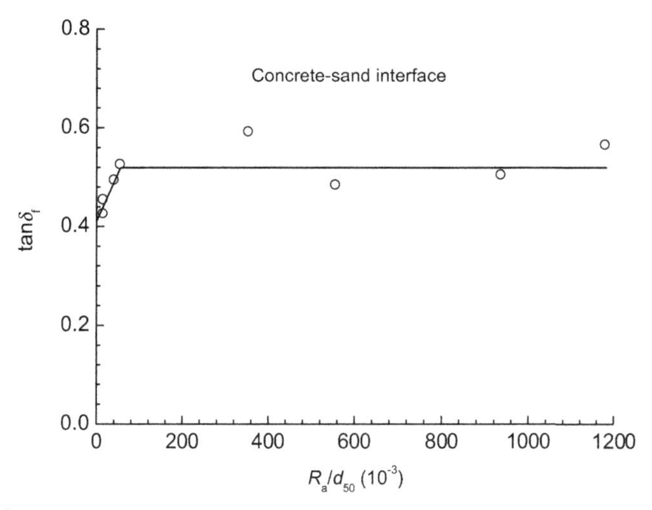

FIGURE 4.2 Relationship between interface roughness R_a normalized by sand d_{50} and $\tan\delta_f$ for silica sand sheared to large displacements against concrete interfaces. *After Barmpopoulos et al. (2009).*

4.3.5 Additional Variables

The calculations also assume the following values for other important parameters.

- The piles' center line average roughnesses are taken as $R_a = 10\,\mu m$ when (1) calculating the radial stress increments developed due to dilatancy on loading to failure and (2) estimating δ_f values.
- Atmospheric pressure is taken as $p_a = 100\,kPa$.
- The diameter of standard CPT devices D_{CPT} is taken as 0.036 m.
- Pile wall thicknesses were input for all open-ended cases.

Chapter | Five

Preliminary Database Analysis of Method Reliabilities

CHAPTER OUTLINE

5.1 TOTAL CAPACITY

A broad summary of the preliminary pile prediction assessments is given in Table 5.1 and Figure 5.1 of the means and CoVs (shown as ±values) found for Q_c/Q_m when the seven methods are evaluated against the ZJU-ICL database. Table 5.1 also adds for reference assessments made by the authors against the tests entered into the original ICP and UWA databases. An additional row is provided in Table 5.1 that gives the statistical summary of how the methods compare when tested against the full ZJU-ICL database without any filtering for test age. The influence of the tests conducted at ages greater than 100 days exceeds that of the early age (<10 day) tests leading to generally lower Q_c/Q_m ratios.

The preliminary results indicate for the filtered ZJU-ICL database:

- Broad agreement with the trends reported by Jardine et al. (2005) and Schneider et al. (2008).
- Mean Q_c/Q_m values spanning from 0.68 to 1.23 overall the cases covered and CoVs from 0.30 to 0.55, with the Main Text API giving consistently higher CoVs than the CPT approaches.
- The "simplified ICP" and "offshore" UWA having significantly lower μ values and larger CoVs than their "full" versions. As shown below, their degrees of fit do not seem to improve as the pile diameter increases.
- No benefit in applying the "simplified ICP" approach in place of the "full" version as it gives an unnecessarily conservative μ and a larger CoV.

A Comprehensive Database of Tests on Axially Loaded Piles Driven in Sand

TABLE 5.1 Summary of ZJU-ICL Assessment of Total Capacity Statistics for API and CPT Methods

	ICP-05		UWA-05				
Database	**Full**	**Simplified**	**Full**	**Offshore**	**Fugro-05**	**NGI-05**	**API**
ICP	0.97 ± 0.35	0.69 ± 0.38	1.00 ± 0.32	0.84 ± 0.38	1.11 ± 0.41	1.16 ± 0.50	0.87 ± 0.66
UWA	0.93 ± 0.34	0.69 ± 0.37	1.00 ± 0.32	0.85 ± 0.38	1.12 ± 0.42	1.19 ± 0.49	0.83 ± 0.63
Age filtered ZJU-ICL data	0.94 ± 0.30	0.68 ± 0.35	1.05 ± 0.35	0.89 ± 0.45	1.20 ± 0.47	1.23 ± 0.48	0.88 ± 0.55
Total ZJU-ICL data	0.92 ± 0.33	0.67 ± 0.41	1.01 ± 0.35	0.85 ± 0.42	1.16 ± 0.55	1.13 ± 0.47	0.85 ± 0.53

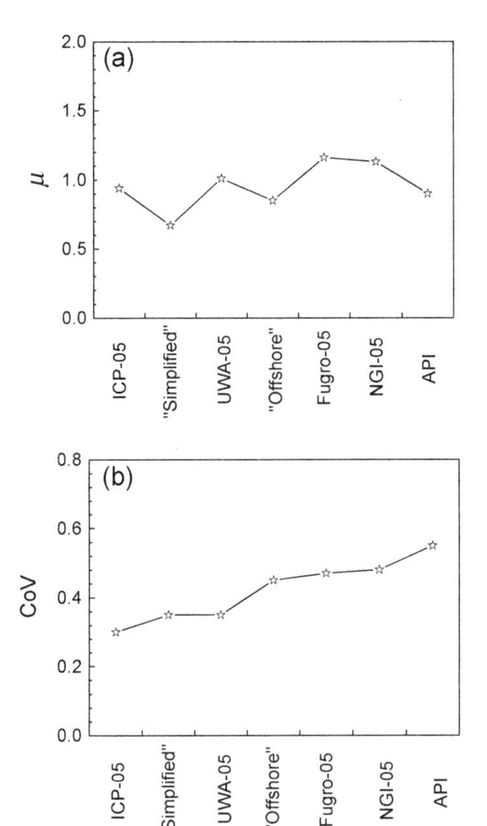

FIGURE 5.1 Statistical values (μ and CoV) of total capacity for design methods based on filtered ZJU-ICL 10–100 day age database.

- The "full" UWA version appearing marginally nonconservative, suggesting that the "offshore" version may be preferable, despite its higher CoV.
- The "full" ICP and UWA methods giving significantly lower CoVs (0.30–0.35 respectively) than the other CPT-based approaches (0.47–0.48) as well as mean Q_c/Q_m values that are closer to unity (0.94–1.05, compared with 1.13–1.16).

Figures 5.2–5.5 present scatter diagrams of Q_c/Q_m in terms of total capacity against pile slenderness ratio L/D, pile diameter D, average q_c along pile shaft $q_{c,avg}$, and average density D_r along shaft for the updated database. As noted by Jardine and Chow (2007) and Schneider et al. (2008), the ICP-05 and UWA-05 CPT-based methods appear to offer the most promise in eliminating the strong skewing produced by the Main Text API method and giving CoV values that are sufficiently low for meaningful reliability analyses to be conducted; see Jardine et al. (2005).

5.2 SHAFT CAPACITY

Shaft and base capacities can only be separated in 44 of the 80 filtered ZJU-ICL cases, including the tension tests. Table 5.2 and Figure 5.6 offer a preliminary statistical assessment of the various shaft capacity procedures, listing as before

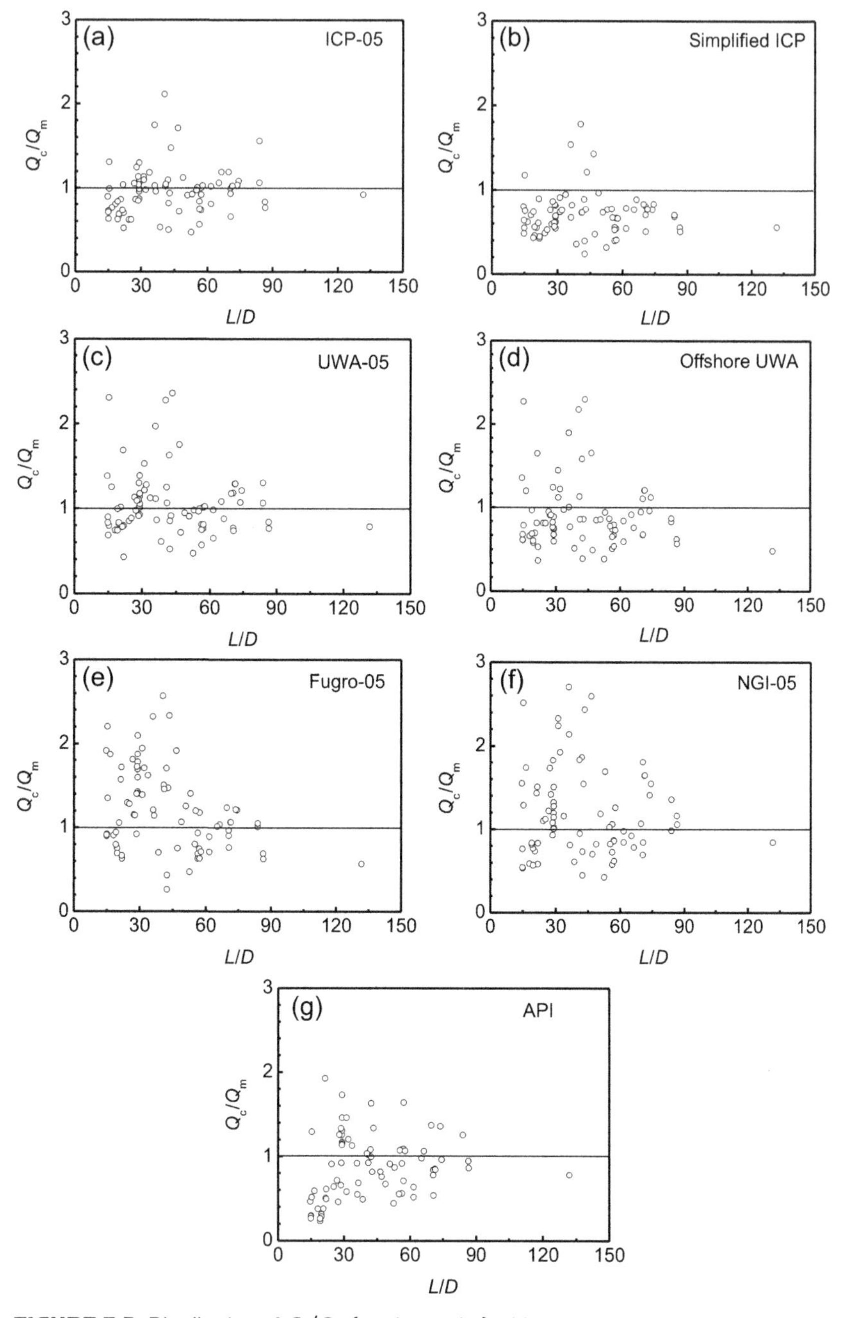

FIGURE 5.2 Distribution of Q_c/Q_m (total capacity) with respect to pile slenderness ratio L/D. (a) ICP-05; (b) "Simplified" ICP-05; (c) UWA-05; (d) "Offshore" UWA-05; (e) Fugro-05; (f) NGI-05; (g) API—tested against filtered ZJU-ICL 10–100 day age data set.

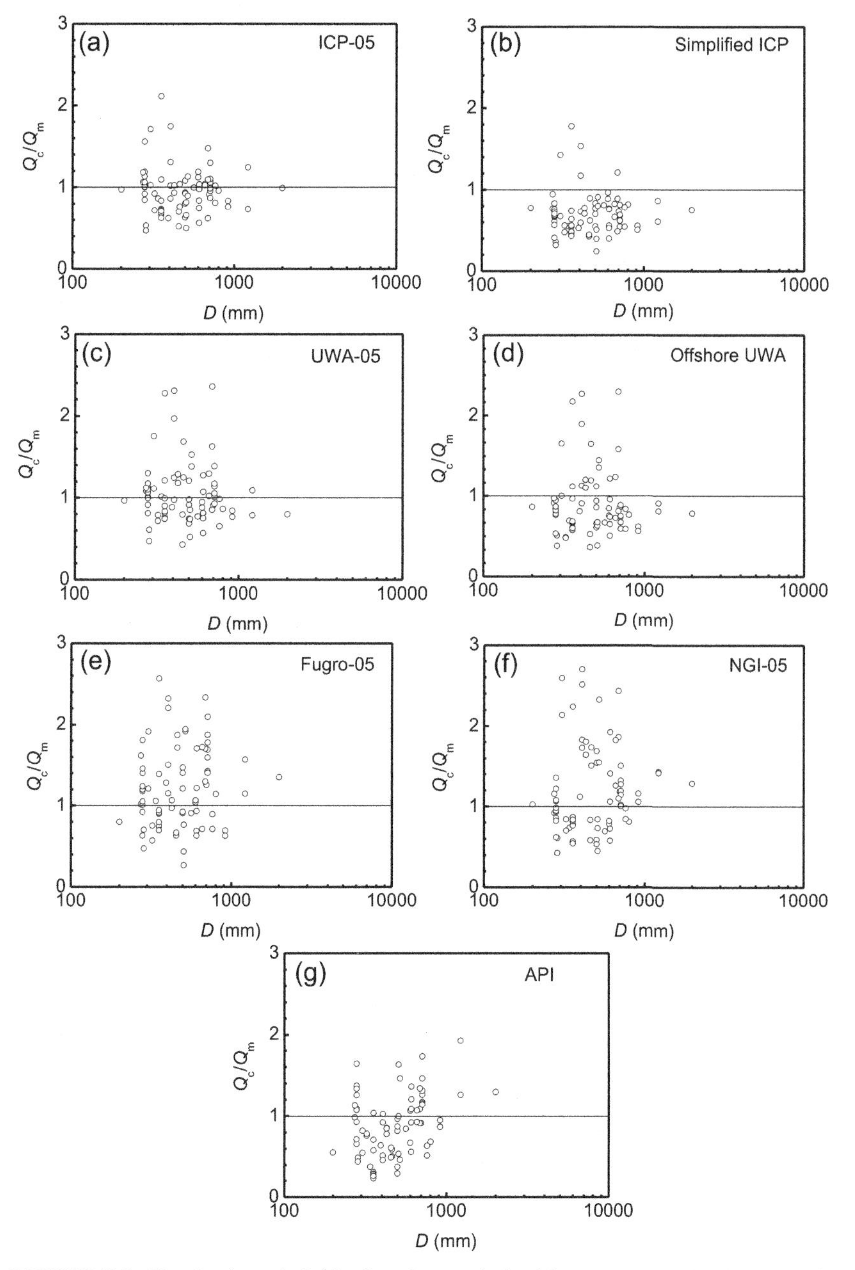

FIGURE 5.3 Distribution of Q_c/Q_m (total capacity) with respect to pile diameter D. (a) ICP-05; (b) "Simplified" ICP-05; (c) UWA-05; (d) "Offshore" UWA-05; (e) Fugro-05; (f) NGI-05; (g) API—tested against filtered ZJU-ICL 10–100 day age data set.

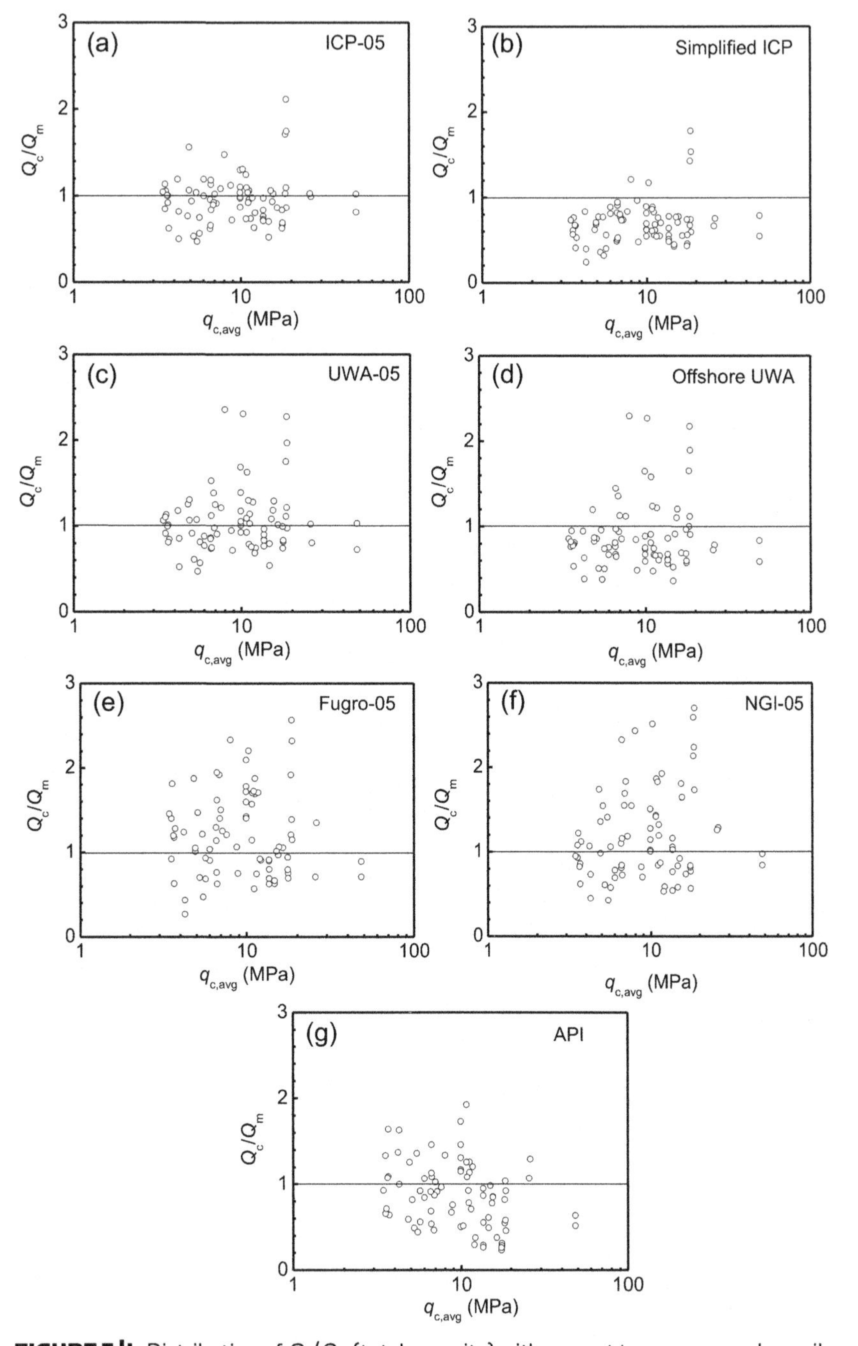

FIGURE 5.4 Distribution of Q_c/Q_m (total capacity) with respect to average q_c along pile shaft, $q_{c,avg}$. (a) ICP-05; (b) "Simplified" ICP-05; (c) UWA-05; (d) "Offshore" UWA-05; (e) Fugro-05; (f) NGI-05; (g) API—tested against filtered ZJU-ICL 10–100 day age data set.

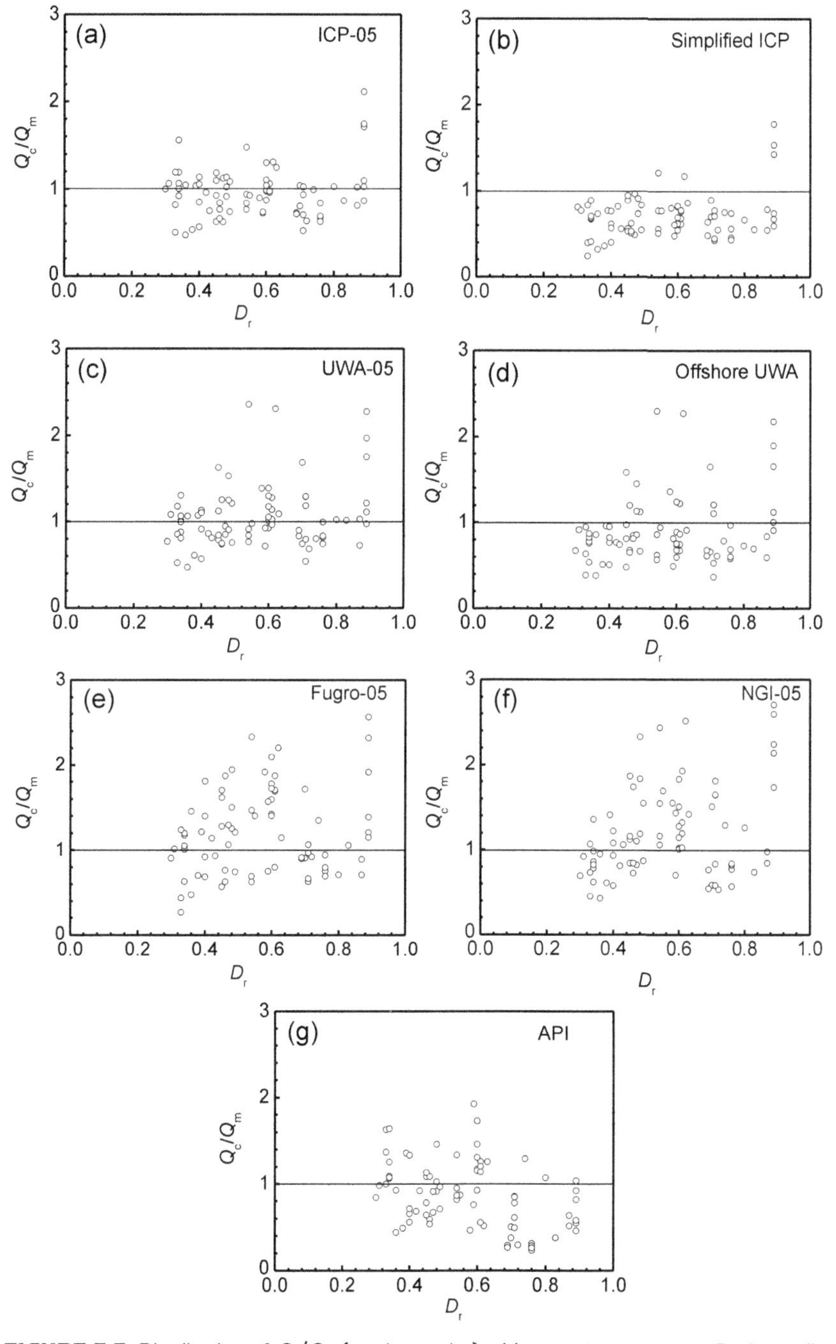

FIGURE 5.5 Distribution of Q_c/Q_m (total capacity) with respect to average D_r along pile shaft. (a) ICP-05; (b) "Simplified" ICP-05; (c) UWA-05; (d) "Offshore" UWA-05; (e) Fugro-05; (f) NGI-05; (g) API—tested against filtered ZJU-ICL 10–100 day age data set.

TABLE 5.2 Summary of ZJU-ICL Assessment of Shaft Capacity Statistics for API and CPT Methods

	ICP-05		UWA-05				
Database	**Full**	**Simplified**	**Full**	**Offshore**	**Fugro-05**	**NGI-05**	**API**
ICP	0.89 ± 0.31	0.55 ± 0.33	0.88 ± 0.27	0.70 ± 0.29	0.82 ± 0.37	0.96 ± 0.53	0.81 ± 0.77
UWA	0.88 ± 0.30	0.54 ± 0.32	0.86 ± 0.27	0.68 ± 0.30	0.79 ± 0.39	0.93 ± 0.51	0.80 ± 0.74
Age filtered ZJU-ICL data	0.86 ± 0.30	0.54 ± 0.39	0.87 ± 0.27	0.66 ± 0.33	0.91 ± 0.41	0.95 ± 0.53	0.84 ± 0.66
Total ZJU-ICL data	0.89 ± 0.35	0.57 ± 0.46	0.92 ± 0.36	0.72 ± 0.40	0.95 ± 0.57	1.00 ± 0.54	0.80 ± 0.59

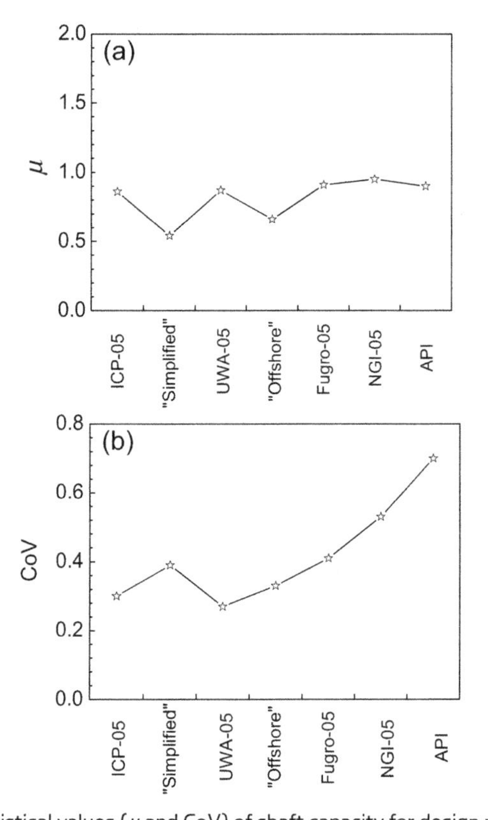

FIGURE 5.6 Statistical values (μ and CoV) of shaft capacity for design methods based on filtered ZJU-ICL 10–100 day age database.

mean and CoV Q_c/Q_m values as assessed from the 10–100 day age ZJU-ICL database where shaft capacity can be reliably determined, leading to the following observations:

- The mean Q_c/Q_m values fall below unity (0.54–0.95) for all the methods, indicating that on average they underestimate the 10–100 day shaft capacities. As shown earlier, the mean ICP predictions tend (on average) to become conservative for piles tested at ages exceeding 10–15 days.
- The "full" ICP and UWA versions give the lowest CoVs and provide better fits to the ZJU-ICL data set than the other methods.
- The "simplified" ICP and "offshore" UWA methods give similar CoVs to their "full" versions, but significantly more conservative means.

Figures 5.7–5.10 show scatter diagrams of Q_c/Q_m in terms of shaft capacity against pile slenderness ratio L/D, pile diameter D, average q_c along pile shaft $q_{c,avg}$, and average density D_r along the shaft for the updated database. The updated database assessment confirms that the "full" versions of ICP-05 and UWA-05 give the least scattered predictions. The "simplified" ICP and "offshore" UWA and API show greater bias with respect to the same parameters.

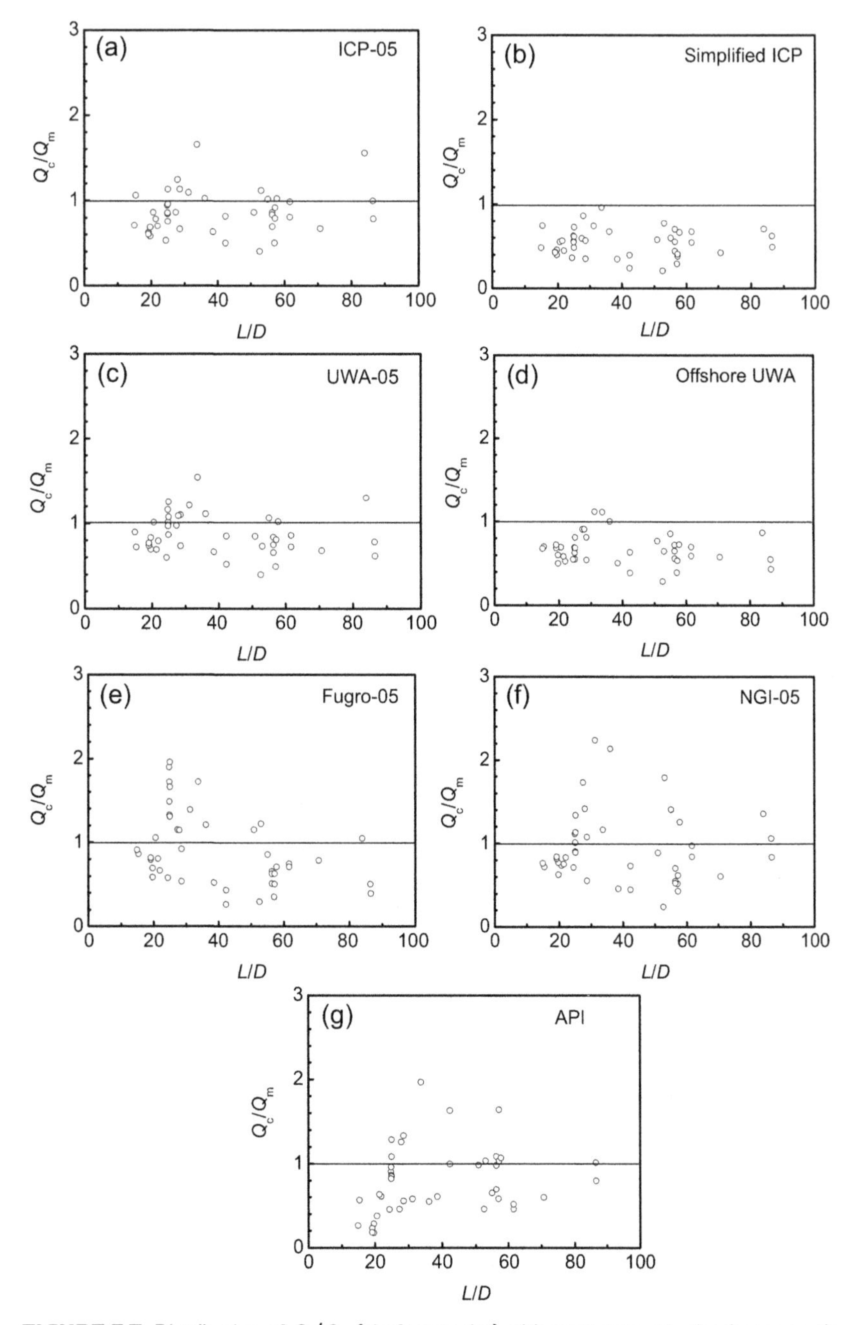

FIGURE 5.7 Distribution of Q_c/Q_m (shaft capacity) with respect to pile slenderness ratio L/D. (a) ICP-05; (b) "Simplified" ICP-05; (c) UWA-05; (d) "Offshore" UWA-05; (e) Fugro-05; (f) NGI-05; (g) API—tested against filtered ZJU-ICL 10–100 day age database.

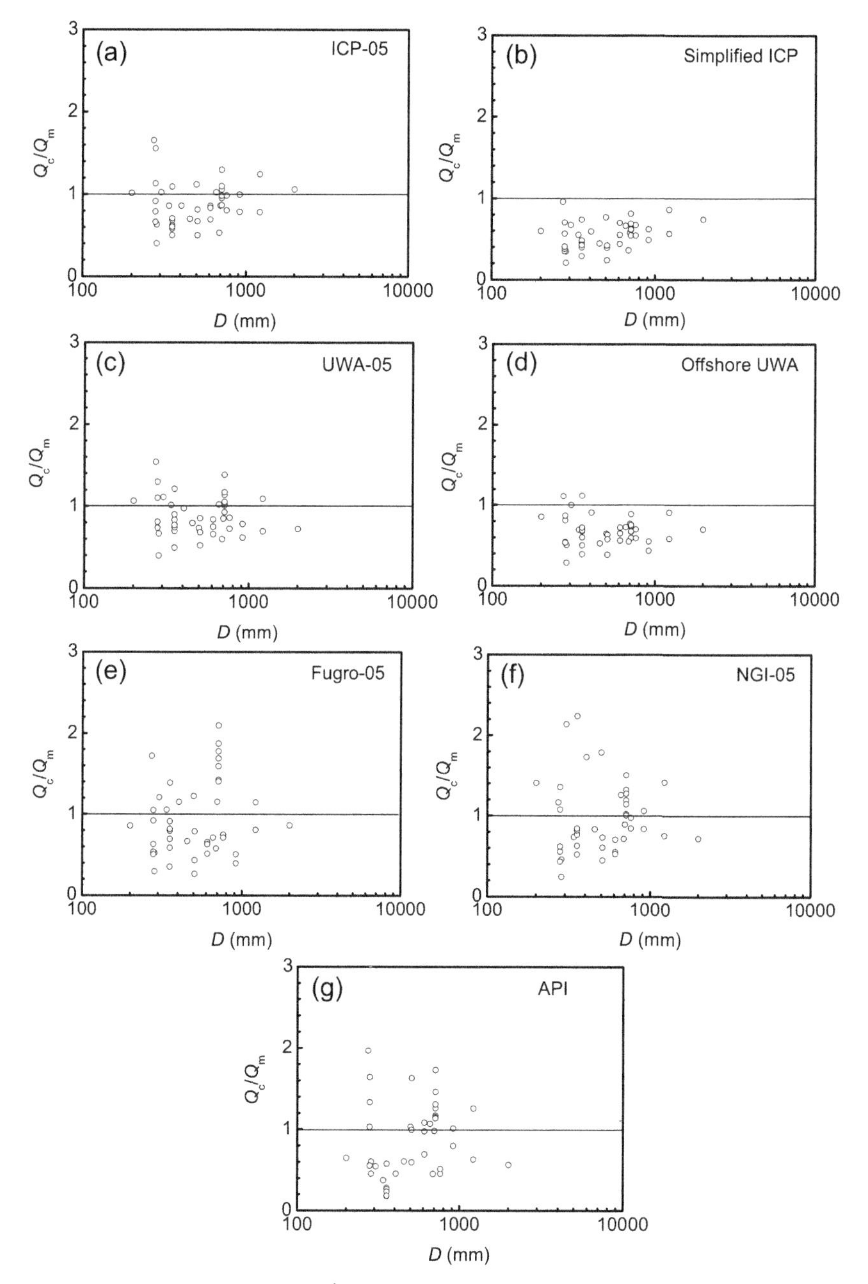

FIGURE 5.8 Distribution of Q_c/Q_m (shaft capacity) with respect to pile diameter D. (a) ICP-05; (b) "Simplified" ICP-05; (c) UWA-05; (d) "Offshore" UWA-05; (e) Fugro-05; (f) NGI-05; (g) API on filtered ZJU-ICL 10–100 day age database.

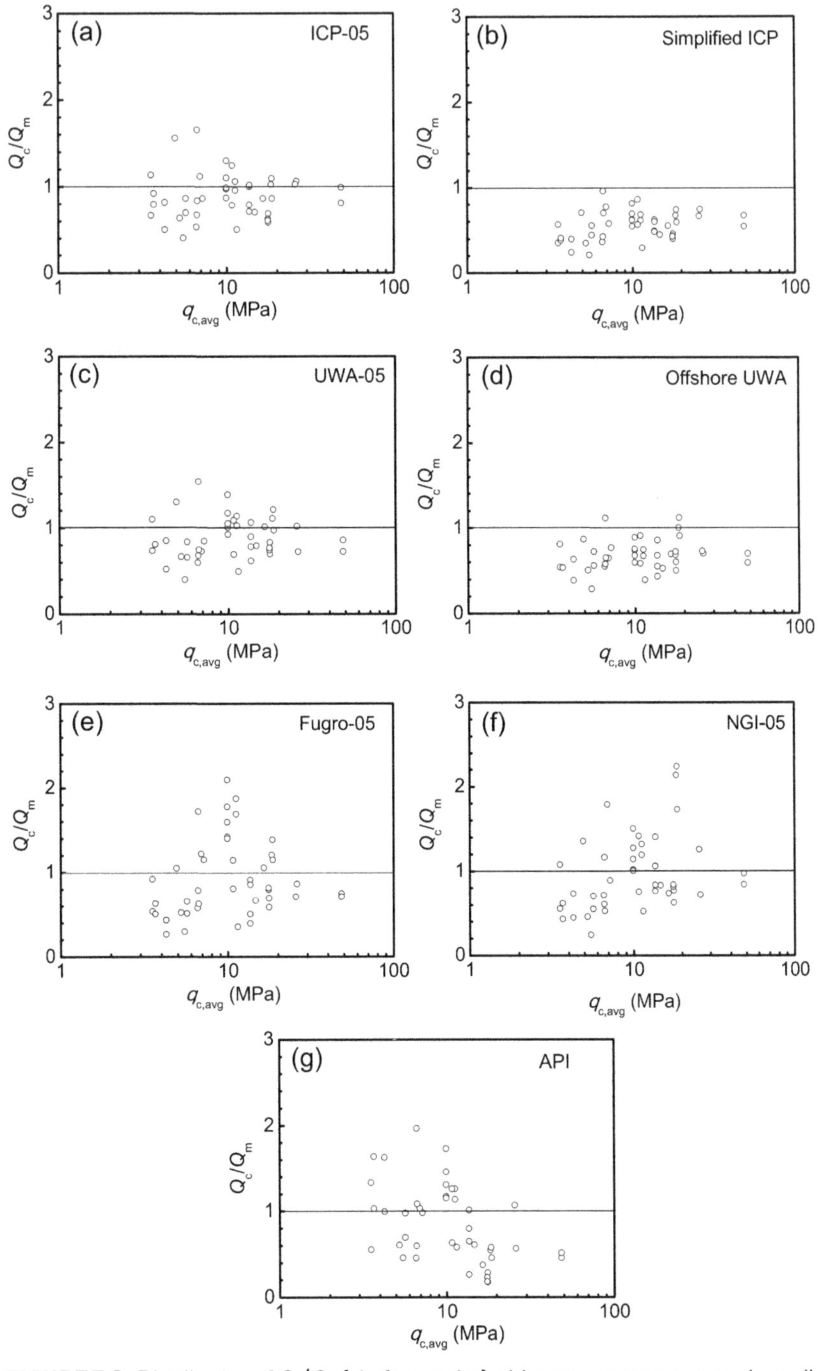

FIGURE 5.9 Distribution of Q_c/Q_m (shaft capacity) with respect to average q_c along pile shaft, $q_{c,avg}$. (a) ICP-05; (b) "Simplified" ICP-05; (c) UWA-05; (d) "Offshore" UWA-05; (e) Fugro-05; (f) NGI-05; (g) API—tested against filtered ZJU-ICL 10–100 day age database.

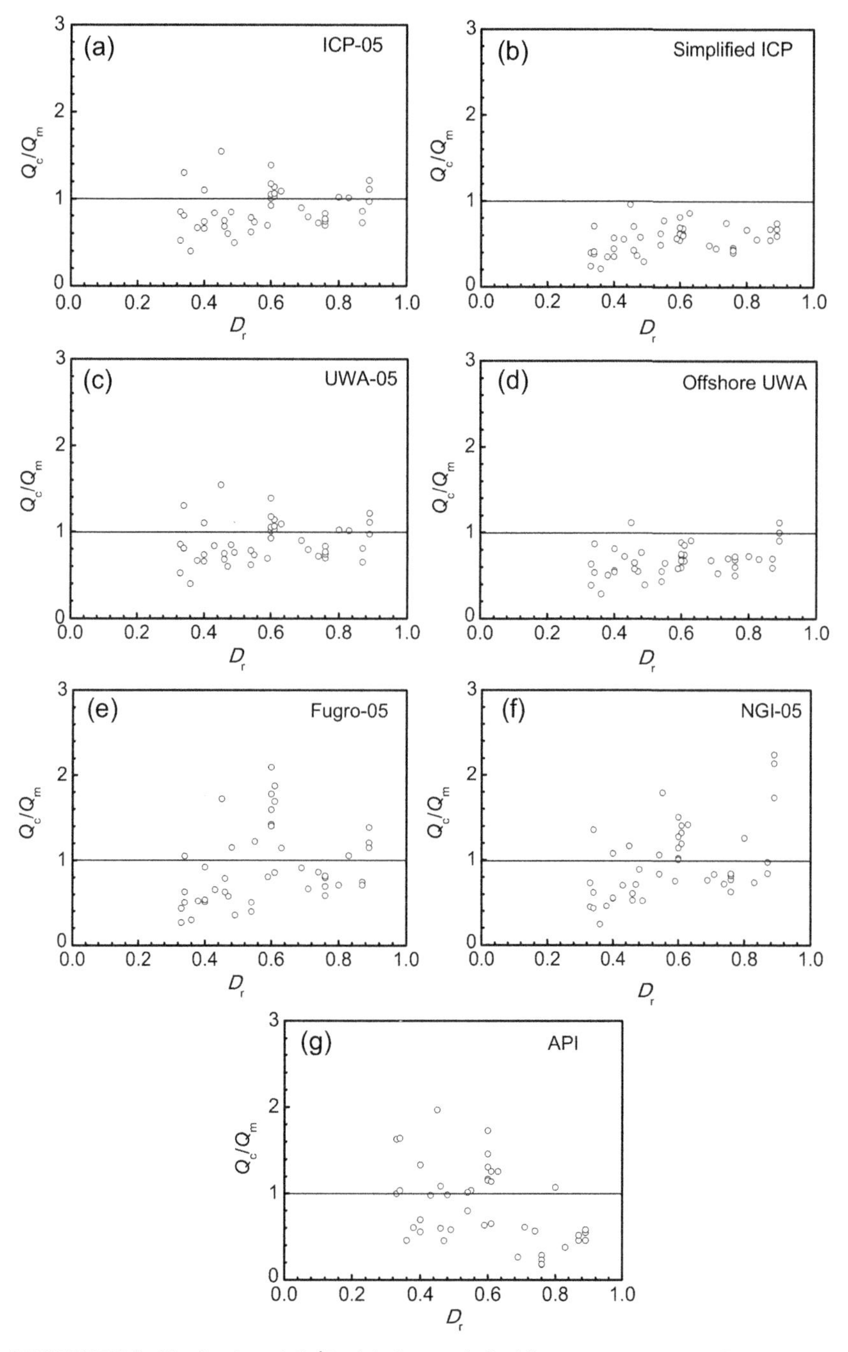

FIGURE 5.10 Distribution of Q_c/Q_m (shaft capacity) with respect to average D_r along pile shaft. (a) ICP-05; (b) "Simplified" ICP-05; (c) UWA-05; (d) "Offshore" UWA-05; (e) Fugro-05; (f) NGI-05; (g) API—tested against filtered ZJU-ICL 10–100 day data set.

TABLE 5.3 Summary of ZJU-ICL Assessment Statistics for Base Capacity; API and CPT Methods and All Base Capacity Data: See Table 2.7 and Appendix for Details of Test Numbers and Data Set Entries

		UWA-05				
Database	**ICP-05**	**Full**	**Offshore**	**Fugro-05**	**NGI-05**	**API**
ICP	0.84±0.38	1.02±0.43	0.93±0.38	1.56±0.48	1.12±0.56	0.95±1.02
UWA	0.82±0.40	1.05±0.41	0.96±0.37	1.54±0.45	1.16±0.56	0.90±0.93
Total ZJU-ICL data	0.84±0.39	1.09±0.42	1.00±0.40	1.63±0.48	1.23±0.59	0.96±0.94

As in Table 5.1, Table 5.2 adds for reference assessments made by the authors with the piles entered into the original ICP, the UWA, and the unfiltered ZJU-UWA databases. As with total capacity, removing the filtering for test age affects the Q_c/Q_m ratios.

5.3 BASE CAPACITY

The updated and filtered database contains only 18 fully credible base capacity measurement cases[1], comprising 13 tests conducted between 10 and 100 days and 5 with unspecified ages that are presumed to fall into the latter age range. This relatively small data set is somewhat sensitive to the individual test outcomes and may not be suitable for statistical treatment. However, as shown by Rimoy (2013) and Rimoy et al. (2015), base capacities show little or no variation with age; so this filter is unnecessarily restrictive. We therefore consider below the filtered 18 cases, plus 11 more that meet all the quality criteria except the age limits, leading to a total data set of 29 cases.

Table 5.3 compares the preliminary statistical summary, listing mean and CoV Q_c/Q_m values of the API Main Text and CPT methods evaluated against the ICP, UWA, and the (29 cases, as described above) ZJU-ICL pile test data sets. Figure 5.11 summarizes the μ and CoV values found with the ZJU-ICL database, while Figures 5.12–5.14 present the distributions of Q_c/Q_m against D and tip q_c and D_r values, respectively. As may be seen, the ICP-05 and UWA-05 methods give the best performance. The ICP appears marginally conservative on average, while the "full UWA" is marginally nonconservative.

[1] Note that two further case histories (with IDs TH52 MAT 2 and RJ PI-3) that pass all the other filtering criteria indicate improbably small end-bearing values that may be associated with problems in estimating residual toe loads and incomplete base capacity development, respectively. These cases have been eliminated from the base capacity database. The first test was accepted into the total capacity data set while the shaft capacity of the second was accepted into the shaft measurement set.

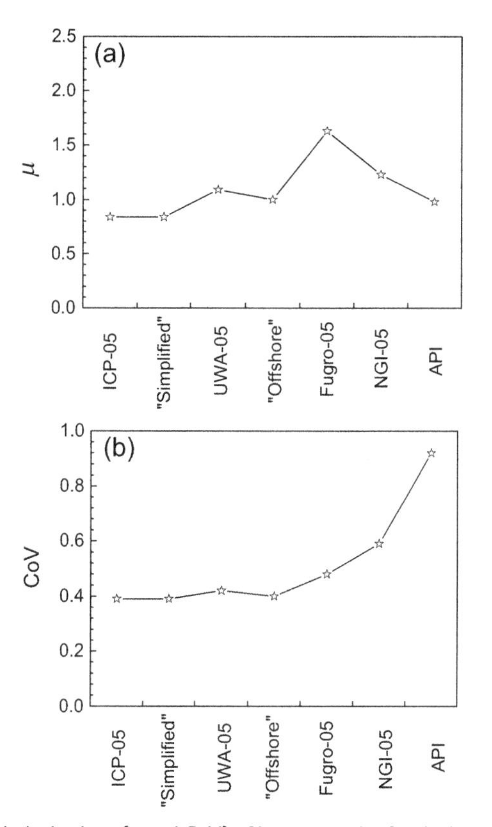

FIGURE 5.11 Statistical values (μ and CoV) of base capacity for design methods based on ZJU-ICL database of 29 tests.

5.4 AGING TRENDS

We consider next the information that can be gleaned from the ZJU-ICL database regarding potential effects of age on the axial capacities of piles driven in sand when subjected to first time testing. Figures 5.15–5.17 present this information as ratio against age of measured-to-calculated capacity (Q_m/Q_c the inverse of Q_c/Q_m) evaluated by various design methods, considering total, shaft, and base capacity, respectively. The piles with unspecified ages are not plotted.

The high degree of scatter displayed by the API predictive method masks the time dependency of total and shaft dependency that is clearer with all of the "CPT-based" method plots. These trends are clearest with the UWA and ICP cases, which have the lowest CoVs. The figures also demonstrate that a significant fraction of the scatter seen in the predictive methods is related to the variable effects of time. Normalizing for such time dependency should reduce the CoVs and affect the degrees of bias significantly. It is also clear from the CPT-based methods' plots in Figure 5.17 that base capacities show little or no variation with time. These findings are consistent with aging trends seen in carefully designed

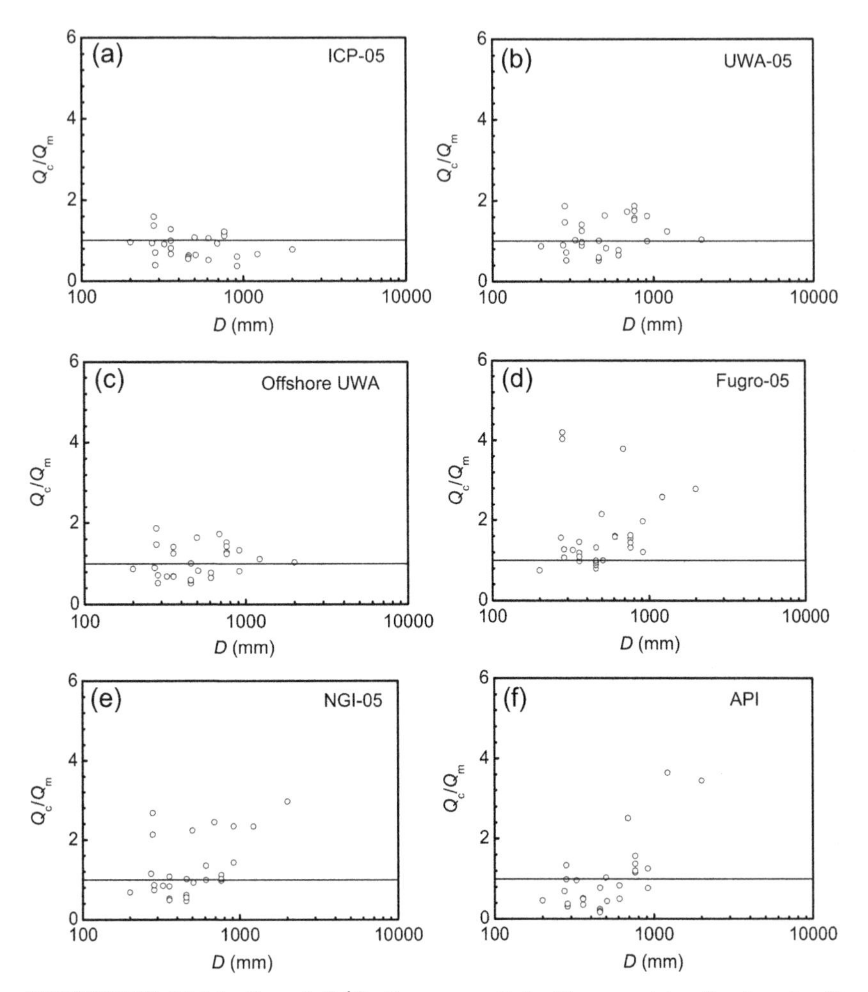

FIGURE 5.12 Distribution of Q_c/Q_m (base capacity) with respect to pile diameter *D*. (a) ICP-05; (b) UWA-05; (c) "Offshore"; (d) Fugro-05; (e) NGI-05; (f) API as evaluated against ZJU-ICL database of 29 tests.

aging tests (see Jardine et al., 2006; Gavin et al., 2013 or Karlsrud et al., 2014) and broader literature reviews, such as those by Rimoy (2013). Shaft capacities build sharply over the weeks and months that follow driving, tending to upper limits after perhaps 1 year, while base capacities remain practically unaffected.

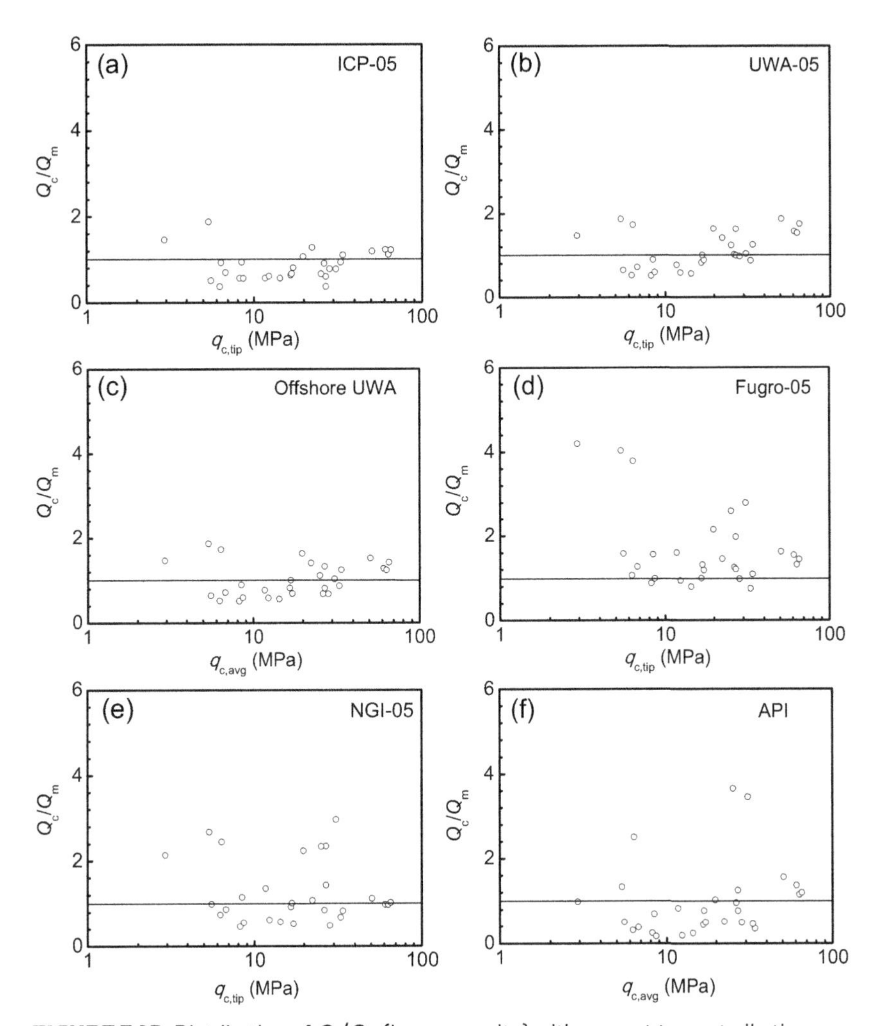

FIGURE 5.13 Distribution of Q_c/Q_m (base capacity) with respect to q_c at pile tip, $q_{c,tip}$. (a) ICP-05; (b) UWA-05; (c) "Offshore"; (d) Fugro-05; (e) NGI-05; (f) API as evaluated against ZJU-ICL database of 29 tests.

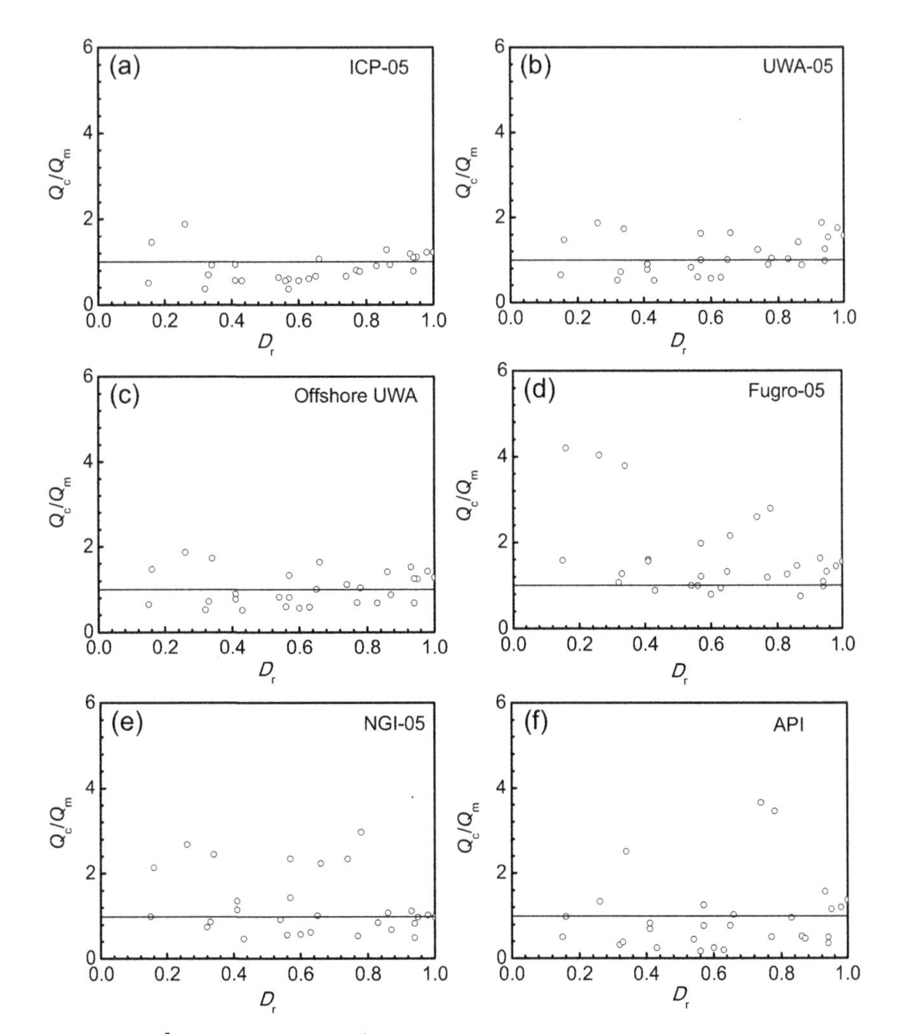

FIGURE 5.14 Distribution of Q_c/Q_m (base capacity) with respect to average D_r at pile tip. (a) ICP-05; (b) UWA-05; (c) "Offshore"; (d) Fugro-05; (e) NGI-05; (f) API as evaluated against ZJU-ICL database of 29 tests.

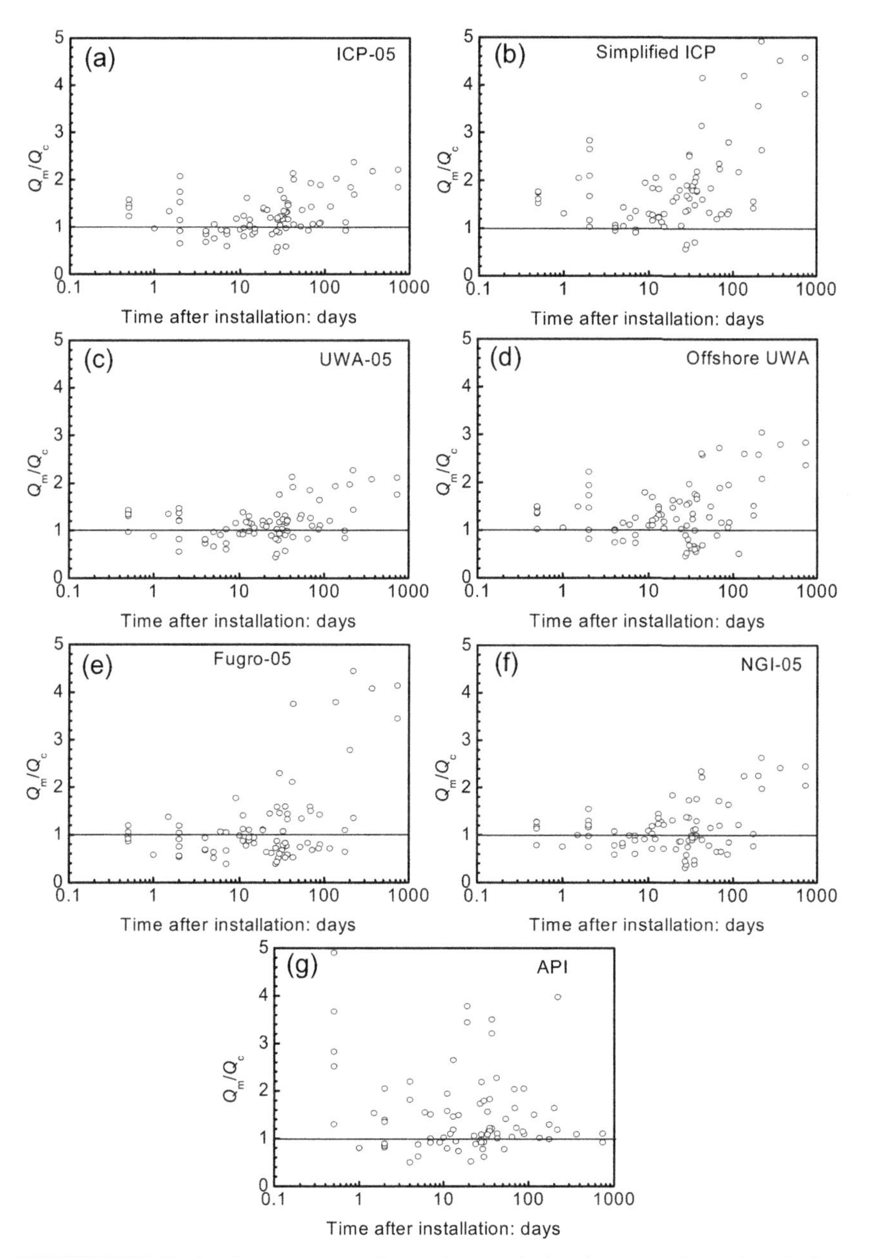

FIGURE 5.15 Ratio of measured total capacity to calculated capacity by various design methods Q_m/Q_c against time after installation (a) ICP-05; (b) "Simplified" ICP-05; (c) UWA-05; (d) "Offshore" UWA-05; (e) Fugro-05; (f) NGI-05; (g) API as evaluated from full ZJU-ICL database.

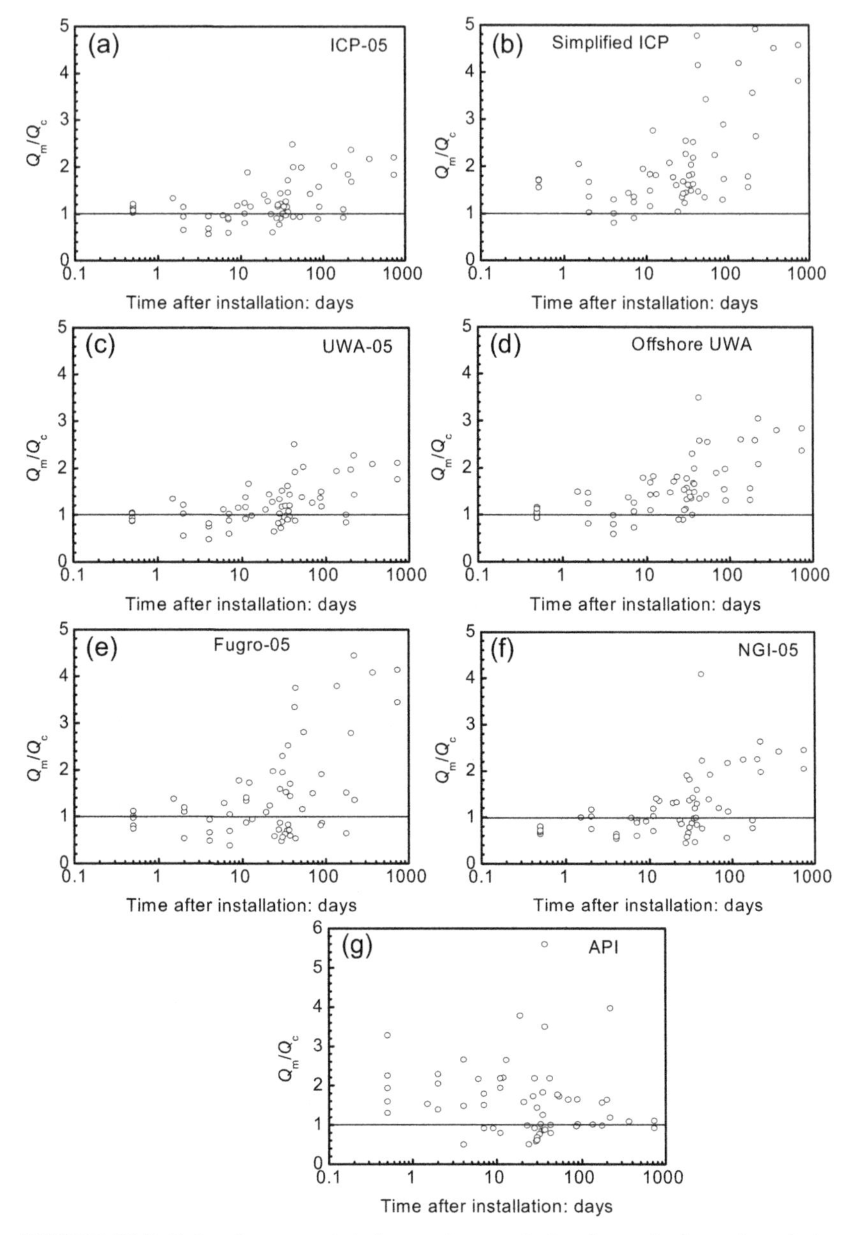

FIGURE 5.16 Ratio of measured shaft capacity to calculated capacity by various design methods Q_m/Q_c against time after installation (a) ICP-05; (b) "Simplified" ICP-05; (c) UWA-05; (d) "Offshore" UWA-05; (e) Fugro-05; (f) NGI-05; (g) API evaluated from full ZJU-ICL database of 53 tests.

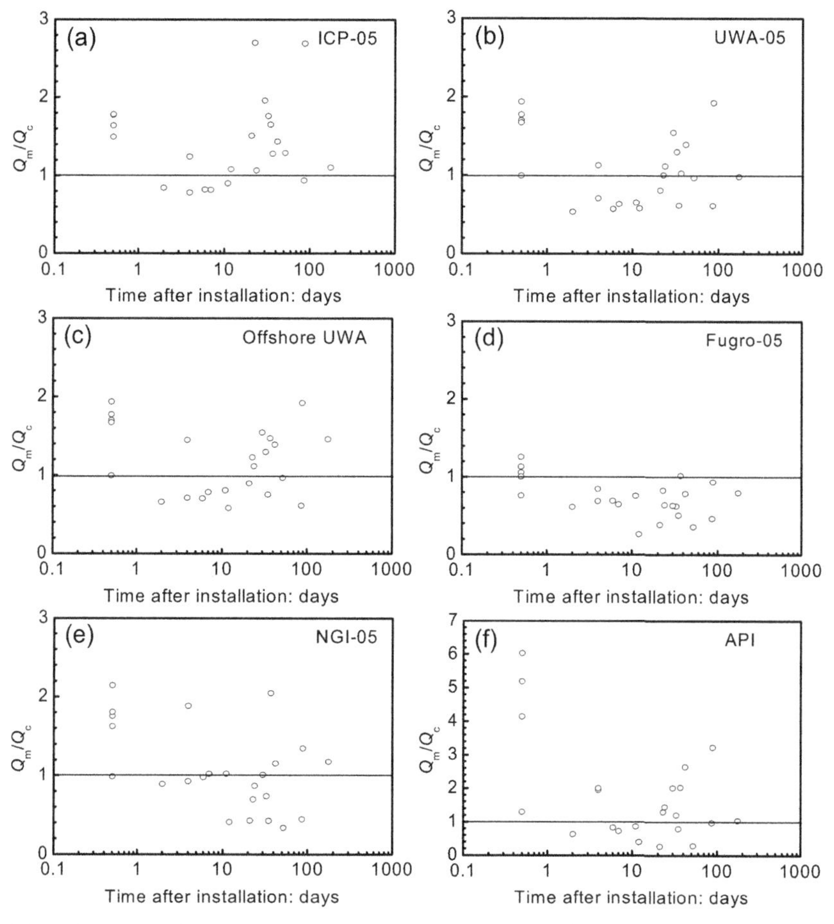

FIGURE 5.17 Ratio of measured base capacity to calculated capacity by various design methods Q_m/Q_c against time after installation (a) ICP-05; (b) UWA-05; (c) "Offshore" UWA-05; (d) Fugro-05; (e) NGI-05; (f) API as evaluated from full ZJU-ICL database where Q_b is known.

Chapter | Six

Summary, Conclusions, and Perspectives

CHAPTER OUTLINE

6.1 SUMMARY

The practical tools available for calculating the capacity and load–displacement behavior of driven piles have, until recently, been relatively unreliable; especially in sand and soft rock ground conditions. Research with highly instrumented piles has led to alternative "CPT-based" sand methods being tabled and included into industrial design guidance; alternative procedures have been put forward for clays and other geomaterials. However, the conclusions as to which methods perform best remain controversial and open to debate.

Database verification plays a vital role in checking the predictive reliability of both existing and new methods of capacity assessment. The scope for moving forwards toward internationally agreed new design procedures depends on updating, upgrading, and augmenting the pile test databases, along with gaining experience on their practical application. The need for improved pile load test databases led ZJU and ICL to carry out the work described in this booklet, which constitutes the first major output from the study and started in 2011. We intend to make our work openly accessible so that other research groups will be able to test and evaluate their design methods independently.

This booklet presents an updated database for piles driven in predominantly silica sands, drawing in new data entries from the team's own projects, the literature and from acknowledged communication with other research groups worldwide. It sets out the background and analytical approach along with a substantial appendix that details each of the assembled tests and its site conditions. The new database adds a 70% increment to the population of high-quality pile load tests that meet the criteria set to test capacity and stiffness design methods. We encourage colleagues to consider submitting further high-quality tests for inclusion into the database so that the value of this freely accessible research resource can continue to grow.

6.2 CONCLUSIONS

The preliminary results obtained from comparisons between the assembled site measurements of axial capacities and predictions from various approaches confirm:

1. The existing Main Text API procedures for piles driven in sand are subject to far larger predictive CoVs than the alternative CPT-based methods.
2. The UWA and ICP procedures appear to give significantly less bias and scatter than the other CPT methods. They appear to offer, for the range of cases covered in the ZJU-ICL database, the best currently available design tools.

A new point to emerge is that the "simplified" ICP method is notably over-conservative for the piles in the ZJU-ICL database. We recommend that future onshore and offshore applications should adopt the "full" formulation set out by Jardine et al. (2005), as endorsed in practical onshore and offshore projects by Williams et al. (1997) and Overy (2007). The preliminary analysis also highlights the significance of pile age on shaft and total capacity. It is clear that this key aspect of behavior has to be addressed in any testing or updating of design methods.

6.3 PERSPECTIVES

Precise prediction remains a challenging task that can only be completed by testing and developing new methods against high-quality test databases. The updated ZJU-ICL database presented in this document provides a useful starting point for making such improvements for piles driven in sand. We trust that it will contribute to sand design method projects that are currently being undertaken by several groups. However, database assessments remain hampered by a lack of high-quality tests, particularly those involving large open-ended pile tests.

The authors will therefore continue to augment and update their electronic database for piles driven in sand and welcome any new test entries that can be submitted that meet the above criteria and data quality levels set out in this booklet.

The ZJU-ICL team will also work to build and develop additional databases that cover axial capacity for piles driven in other geomaterials, starting with clays, and provide updated guidance on driven pile load–displacement behavior.

ZJU-ICL Database

PREFACE

This booklet comprises the first major output from the study. It presents an updated database for piles driven in predominantly silica sands, drawing in new data entries from the team's own projects, the literature and from acknowledged communication with other research groups worldwide. We intend that our openly accessible database will enable other research groups to test and evaluate their design methods independently.

The appendix herein details each of the assembled tests and its site conditions collected in the ZJU-ICL database. In total, 52 sites and 116 pile load tests are involved in this appendix. The appendix comprises of three parts: new data entries, accepted ICP data, and accepted UWA data. The site information and details of pile load tests will be given separately (Tables A.1 and A.2). For each site, a CPT q_c profile will be given along with the information of site name, location, soil type, water table depth, pile type, number of pile load tests, and some comments. The pile load test information includes the load–displacement curve if available, pile type and material, pile dimensional parameters, installation method, setup time, loading mode, failure capacities, etc. The assessment performance of the API Main Text approach and four CPT-based procedures are also summarized. A full reference to each case is also provided to allow the readers to browse the data directly.

NOTATION

q_c	CPT tip resistance
Q_b	Base capacity
Q_c	Calculated pile axial capacity
Q_m	Capacities used for database assessment, referring to the maximum capacities measured in field tests or capacities measured at a pile tip settlement of 10% of the pile diameter in compression
$Q_{max\text{-}measured}$	Maximum capacities measured in field tests
Q_s	Shaft capacity
API	American Petroleum Institute
CPT	Cone penetration test
CPTU	Piezocone penetration test
ECPT	Electrical cone penetration test

ICP	Imperial College pile
IFR	Incremental filling ratio
PSD	Particle size distribution
MCPT	Mechanical cone penetration test
NGI	Norwegian Geotechnical Institute
PHC	Prestressed high-strength concrete piles
UWA	University of Western Australia
ZJU-ICL	Zhejiang University-Imperial College London

Table A.1 Summary of Site Information in ZJU-ICL Database

Site ID	Site name and location	Pile type		Pile material		Load test type		References	Page
		Open	Closed	Concrete	Steel	Compression	Tension		
1	Wuhu, K24, China	3		3		3		Yang et al. (2015a)	74–77
2	Wuhu, K34, China	1		1		1		Yang et al. (2015a)	78–79
3	Wuhu, K27, China	1		1		1		Yang et al. (2015a)	80–81
4	Rio de Janeiro, Brazil		2	2		2		Tsuha (2012)	82–84
5	Rio de Janeiro, Brazil		2	2		2		Tsuha (2012)	85–87
6	Blessington Dublin, Ireland	3			3		3	Gavin et al. (2013)	88–91
7	Horstwalde, Germany	8			8		8	Rücker et al. (2013)	92–101
8	British Columbia, Canada		1		1	1		Naesgaard et al. (2012)	102–103
9	Hampton Virginia, USA		1	1		1		Pando et al. (2003)	104–105
10	Rotterdam Harbor, The Netherlands		3	3		3		de Gijt et al. (1995)	106–109
11	Waddinxveen Site, The Netherlands		1	1		1		Hölscher (2009)	110–111
12	Mobile Bay, AL, USA	1			1	1		Mayne (2013)	112–113
13	Mobile Bay, AL, USA	1			1	1		Mayne (2013)	114–115
14	ABEF Foundation, Brazil	1			1	1		Mayne (2013)	116–117
15	ABEF Foundation, Brazil	1			1	1		Mayne (2013)	118–119

Continued

Table A.1 Summary of Site Information in ZJU-ICL Database—continued

Site ID	Site name and location	Pile type		Pile material		Load test type		References	Page
		Open	Closed	Concrete	Steel	Compression	Tension		
16	Apalachicola River, USA		1	1		1		Mayne (2013)	120–121
17	Los Angeles, CA Site, USA		1	1		1		Mayne (2013)	122–123
18	MS Smith, USA		1	1		1		Mayne (2013)	124–125
19	MS Desota, USA		1	1		1		Mayne (2013)	126–127
20	MS Harrison, USA		1	1		1		Mayne (2013)	128–129
21	Washington MS, USA		1	1		1		Mayne (2013)	130–131
22	Washington MS, USA		1	1		1		Mayne (2013)	132–133
23	Washington MS, USA		1	1		1		Mayne (2013)	134–135
24	Larvik, Norway	7			7		7	Karlsrud et al. (2014)	136–143
25	Jackson Country, USA		1		1	1		Mayne and Elhakim (2002)	144–145
26	Lafayette Bridge, USA		1		1	1		Komurka and Grauvogl-Graham (2010)	146–147
27	Ogeechee River, USA		6	1	5	5	1	Vesic (1970)	148–154
28	Drammen, Norway		3	3		7	3	Gregersen et al. (1973)	155–165
29	Hoogzand, The Netherlands	2	1	3		3	3	Beringen et al. (1979)	166–172
30	Hunter's Point, USA		1		1	1		Briaud et al. (1989a)	173–174
31	Akasaka, Tokyo, Japan		1		1	1		BCP-committee (1971)	175–176

32	Hound Point, Scotland	3			3	1	2	Williams et al. (1997)	177–180
33	Leman BD, North Sea	1			1		1	Jardine et al. (1998)	181–182
34	Baghdad University, Iraq		2	2		2	1	Altaee et al. (1992)	183–186
35	Dunkirk, France	2			2		2	Chow (1997)	187–189
36	Dunkirk, France	2			2	1	2	Jardine et al. (2006)	190–193
37	Euripides, The Netherlands	3			3	3	3	Kolk et al. (2005b)	194–200
38	Euripides, The Netherlands	1			1	1	1	Kolk et al. (2005b)	201–203
39	Locks and Dam, USA		6		6	3	3	Briaud et al. (1989b)	204–210
40	Tokyo Bay, Japan	1			1	1		Shioi et al. (1992)	211–212
41	Hsin-Ta, Taiwan, China		1		1	1		Yen et al. (1989)	213–214
42	Hsin-Ta, Taiwan, China		1		1		1	Yen et al. (1989)	215–216
43	Hsin-Ta, Taiwan, China		1		1	1		Yen et al. (1989)	217–218
44	Drammen, Norway	1			1	1		Tveldt and Fredriksen (2003)	219–220
45	Drammen, Norway	1			1	2		Tveldt and Fredriksen (2003)	221–223
46	Shanghai, China	2			2	2		Pump et al. (1998)	224–226
47	Cimarron River, USA		2	1	1	2		Nevels and Snethen (1994)	227–229
48	Jonkoping, Sweden		3	3		3		Jendeby et al. (1994)	230–233
49	Fittja Straits, Sweden		1	1		2		Axelsson (2000)	234–236
50	Sermide, Italy		1		1	1		Appendino (1981)	237–238
51	Pigeon River, USA		1		1	1		Paik et al. (2003)	239–240
52	Pigeon River, USA	1			1	1		Paik et al. (2003)	241–242

Table A.2 Glossary of Terms in Tables A.3–A.5

Column	Description
Test ID	ID number for case in ZJU-ICL database
Site	Site name shown in source
Pile No.	Pile ID number
Pile material	Material from which pile was made; C = concrete and S = steel
Pile shape	Exterior shape of pile; C = circular; S = square; O = octagonal
B or D	Outer width of square pile or octagonal piles or diameter of circular piles
t	Wall thickness for open-ended pile
z_{tip}	Tip depth of pile
Test type	Compression or tension test; C = compression and T = tension
Age	Time of load testing after pile driven
IFR	Incremental filling ratio of open-ended pile
δ_f	Interface friction angle with default value = 29°

Table A.3 Summary of New Entries in ZJU-ICL Database

Test ID	Site name	Pile No.	Pile material	Pile shape	*B* or *D* (mm)	*t* (mm)	z_{tip} (m)	Test type	Age (days)	Average IFR	Interface friction angle δ_f	Page
001	Wuhu	K24-1	C	C	600	130	33	C	5	0.74	Estimated by PSD	75
002	Wuhu	K24-2	C	C	600	130	39.8	C	14	0.74	Estimated by PSD	76
003	Wuhu	K24-3	C	S	500	127	39.8	C	13	0.73	Estimated by PSD	77
004	Wuhu	K34-1	C	C	600	130	29.3	C	15	0.82	Estimated by PSD	79
005	Wuhu	K27-1	C	C	800	130	29.2	C	13	0.74	Estimated by PSD	81
006	Rio de Janeiro	PI-1	C	C	500	–	37.2	C	64	–	Default value	83
007	Rio de Janeiro	PI-2a	C	C	500	–	21.4	C	72	–	Default value	84
008	Rio de Janeiro	PI-3	C	C	700	–	35.6	C	89	–	Default value	86
009	Rio de Janeiro	PI-4	C	C	500	–	26.5	C	86	–	Default value	87
010	Dublin	S2	S	C	340	14	7	T	2	0.73	From ring shear test	89
011	Dublin	S3	S	C	340	14	7	T	13	0.73	From ring shear test	90
012	Dublin	S5	S	C	340	14	7	T	220	0.73	From ring shear test	91
013	Horstwalde	P2B	S	C	711	12.5	17.61	T	43	0.86	Default value	94
014	Horstwalde	P2D	S	C	711	25	17.69	T	34	0.85	Default value	95
015	Horstwalde	P5B	S	C	711	12.5	17.71	T	36	0.86	Default value	96
016	Horstwalde	P5D	S	C	711	12.5	17.76	T	29	0.86	Default value	97
017	Horstwalde	P4B	S	C	711	12.5	17.67	T	37	0.86	Default value	98

Continued

Table A.3 Summary of New Entries in ZJU-ICL Database—continued

Test ID	Site name	Pile No.	Pile material	Pile shape	*B* or *D* (mm)	*t* (mm)	z_{tip} (m)	Test type	Age (days)	Average IFR	Interface friction angle δ_f	Page
018	Horstwalde	P4D	S	C	711	12.5	17.66	T	32	0.86	Default value	99
019	Horstwalde	P3B	S	C	711	12.5	17.63	T	116	0.86	Default value	100
020	Horstwalde	P3D	S	C	711	12.5	17.74	T	30	0.86	Default value	101
021	Columbia	P1	S	C	610	–	45	C	15	–	Default value	103
022	Hampton River	P1	C	S	610	–	16.8	C	12	–	Default value	105
023	Rotterdam	P6	C	S	380	–	30.6	C	–	–	Default value	107
024	Rotterdam	P8	C	S	380	–	30.3	C	–	–	Default value	108
025	Rotterdam	P10	C	S	380	–	30.7	C	–	–	Default value	109
026	Waddinxveen	P2	C	S	350	–	10	C	33	–	Default value	111
027	Mobile Bay	AL 1	S	C	324	25.4	15.2	C	–	0.71	Default value	113
028	Mobile Bay	AL 2	S	C	324	25.4	42.7	C	–	0.71	Default value	115
029	ABEF Foundation	7	C	C	500	90	9.0	C	–	0.73	Default value	117
030	ABEF Foundation	8	C	C	500	90	7.5	C	–	0.73	Default value	119
031	Apalachicola	BR 1	C	S	610	–	29.9	C	–	–	Default value	121
032	Los Angeles	CA	C	S	610	–	29	C	–	–	Default value	123
033	MS Smith	1045	C	S	410	–	10.2	C	–	–	Default value	125
034	MS Desoto	2108	C	S	460	–	7.6	C	–	–	Default value	127

035	MS Harrison	3028	C	S	460	–	16.2	C	–	–	Default value	129
036	Washington	3118A	C	S	410	–	7.6	C	–	–	Default value	131
037	Washington	3123B	C	S	360	–	16.6	C	–	–	Default value	133
038	Washington	3142A	C	S	360	–	6.2	C	–	–	Default value	135
039	Larvik Site	L1	S	C	508	6.3	21.5	T	43	0.80	Default value	137
040	Larvik Site	L2	S	C	508	6.3	21.5	T	135	0.80	Default value	138
041	Larvik Site	L3	S	C	508	6.3	21.5	T	218	0.80	Default value	139
042	Larvik Site	L4	S	C	508	6.3	21.5	T	365	0.80	Default value	140
043	Larvik Site	L5	S	C	508	6.3	21.5	T	730	0.80	Default value	141
044	Larvik Site	L6	S	C	508	6.3	21.5	T	730	0.80	Default value	142
045	Larvik Site	L7	S	C	508	6.3	21.5	T	30	0.80	Default value	143
046	Jackson County	JCEPF 2	S	C	273	–	17.8	C	10	–	Default value	145
047	Lafayette BRG	MAT2	S	C	356	–	20.3	C	54	–	Default value	147
048	Fittja Straits	D-1	C	S	235	–	13.0	C	1	–	Estimated by PSD	236

Table A.4 Summary of ICP Data Entries in ZJU-ICL Database

Test ID	Site name	Pile No.	Pile material	Pile shape	*B* or *D* (mm)	*t* (mm)	z_{tip} (m)	Test type	Age (days)	Average IFR	Interface friction angle δ_f	Page
001	Ogeechee River	H-2	C	S	406	–	15.2	C	0.5	–	Estimated by PSD	149
002	Ogeechee River	H-12	S	C	457	–	6.1	C	0.5	–	Estimated by PSD	150
003	Ogeechee River	H-13	S	C	457	–	8.9	C	0.5	–	Estimated by PSD	151
004	Ogeechee River	H-14	S	C	457	–	12	C	0.5	–	Estimated by PSD	152
005	Ogeechee River	H-15	S	C	457	–	15	C	0.5	–	Estimated by PSD	153
006	Ogeechee River	H-16	S	C	457	–	15	T	1.5	–	Estimated by PSD	154
007	Drammen	A	C	C	280	–	8	C	–	–	Default value	156
008	Drammen	D/A	C	C	280	–	16	C	–	–	Default value	157
009	Drammen	E-7.5	C	C	280	–	7.5	C	–	–	Default value	158
010	Drammen	E-11.5	C	C	280	–	11.5	C	–	–	Default value	159
011	Drammen	E-15.5	C	C	280	–	15.5	C	–	–	Default value	160
012	Drammen	E-19.5	C	C	280	–	19.5	C	–	–	Default value	161
013	Drammen	E-23.5	C	C	280	–	23.5	C	–	–	Default value	162
014	Drammen	A(T)	C	C	280	–	8	T	–	–	Default value	163
015	Drammen	D/A(T)	C	C	280	–	16	T	–	–	Default value	164
016	Drammen	E-(T)	C	C	280	–	23.5	T	–	–	Default value	165
017	Hoogzand	1-C	S	C	356	16	7	C	37	0.66	Estimated by PSD	167

018	Hoogzand	1-T	S	C	356	16	7	T	37	0.66	Estimated by PSD	168
019	Hoogzand	3-C	S	C	356	20	5.3	C	19	0.77	Estimated by PSD	169
020	Hoogzand	3-T	S	C	356	20	5.3	T	19	0.77	Estimated by PSD	170
021	Hoogzand	2-C	S	C	356	–	6.8	C	–	–	Estimated by PSD	171
022	Hoogzand	2-T	S	C	356	–	6.8	T	–	–	Estimated by PSD	172
023	Hunter's Point	S	S	C	273	–	9.2	C	24	–	Default value	174
024	Akasaka	6C	S	C	200	–	11	C	–	–	Estimated by PSD	176
025	Hound Point	P(0)-C	S	C	1220	24.2	26	C	21	0.95	Default value	178
026	Hound Point	P(0)-T1	S	C	1220	24.2	34	T	11	0.95	Default value	179
027	Hound Point	P(0)-T2	S	C	1220	24.2	41	T	4	0.95	Default value	180
028	Lemen	BD	S	C	660	19	38.1	T	–	0.84	Default value	182
029	Baghdad	P1-C	C	S	253	–	11	C	88	–	Estimated by PSD	184
030	Baghdad	P1-T	C	S	253	–	11	T	200	–	Estimated by PSD	185
031	Baghdad	P2-C	C	S	253	–	15	C	42	–	Estimated by PSD	186
032	Dunkirk	CL-T	S	C	324	12.7	11.3	T	175	0.72	From ring shear test	188
033	Dunkirk	CS-T	S	C	324	19.1	11.3	T	187	0.72	From ring shear test	189
034	Dunkirk	R1-T	S	C	457	13.5	19.3	T	9	0.78	From ring shear test	191
035	Dunkirk	C1-C	S	C	457	13.5	10	C	68	0.78	From ring shear test	192
036	Dunkirk	C1-T	S	C	457	13.5	10	T	69	0.78	From ring shear test	193
037	Euripides	Ia	S	C	763	35.6	30.5	C	7	0.99	From ring shear test	195
038	Euripides	Ib	S	C	763	35.6	38.7	C	2	0.97	From ring shear test	196

Continued

Table A.4 Summary of ICP Data Entries in ZJU-ICL Database—continued

Test ID	Site name	Pile No.	Pile material	Pile shape	*B* or *D* (mm)	*t* (mm)	z_{tip} (m)	Test type	Age (days)	Average IFR	Interface friction angle δ_f	Page
039	Euripides	Ic	S	C	763	35.6	47	C	11	0.96	From ring shear test	197
040	Euripides	Ia-T	S	C	763	35.6	30.5	T	7	0.99	From ring shear test	198
041	Euripides	Ib-T	S	C	763	35.6	38.7	T	2	0.97	From ring shear test	199
042	Euripides	Ic-T	S	C	763	35.6	47	T	11	0.96	From ring shear test	200
043	Euripides	II	S	C	763	35.6	46.7	C	6	0.95	From ring shear test	202
044	Euripides	II-T	S	C	763	35.6	46.7	T	7	0.95	From ring shear test	203
045	Lock and Dam 26	3-1	S	C	305	–	14.2	C	35	–	Estimated by PSD	205
046	Lock and Dam 26	3-4	S	C	356	–	14.4	C	27	–	Estimated by PSD	206
047	Lock and Dam 26	3-7	S	C	406	–	14.6	C	28	–	Estimated by PSD	207
048	Lock and Dam 26	3-2	S	C	305	–	11	T	35	–	Estimated by PSD	208
049	Lock and Dam 26	3-5	S	C	356	–	11.1	T	27	–	Estimated by PSD	209
050	Lock and Dam 26	3-8	S	C	406	–	11.1	T	28	–	Estimated by PSD	210
051	Tokyo Bay	TP	S	C	2000	34	30.6	C	52	1.00	Default value	212
052	Hsin-Ta	TP 4	S	C	609	–	34.3	C	33	–	Default value	214
053	Hsin-Ta	TP 5	S	C	609	–	34.3	T	28	–	Default value	216
054	Hsin-Ta	TP 6	S	C	609	–	34.3	C	30	–	Default value	218

Table A.5 Summary of UWA Data Entries in ZJU-ICL Database

Test ID	Site name	Pile No.	Pile material	Pile shape	*B* or *D* (mm)	*t* (mm)	z_{tip} (m)	Test type	Age (days)	Average IFR	Interface friction angle δ_f	Page
001	Drammen	16-P1-11	S	C	813	12.5	11	C	2	0.88	Default value	220
002	Drammen	25-P2-15	S	C	813	12.5	15	C	2	0.88	Default value	222
003	Drammen	25-P2-25	S	C	813	12.5	25	C	2	0.88	Default value	223
004	Shanghai	ST-1	S	C	914	20	79	C	23	0.80	Default value	225
005	Shanghai	ST-2	S	C	914	20	79.1	C	35	0.85	Default value	226
006	Cimarron River	P1	S	C	660	–	19	C	–	–	Default value	228
007	Cimarron River	P2	C	O	610	–	19.5	C	–	–	Default value	229
008	Jonkoping	P23	C	S	235	–	16.8	C	>1	–	Default value	231
009	Jonkoping	P25	C	S	235	–	17.8	C	<1	–	Default value	232
010	Jonkoping	P26	C	S	275	–	16.2	C	>1	–	Default value	233
011	Fittja Straits	D-5	C	S	235	–	12.8	C	5	–	Estimated by PSD	235
012	Sermide	S	S	C	508	–	35.9	C	–	–	Default value	238
013	Pigeon Creek	1	S	C	356	–	6.9	C	4	–	Default value	240
014	Pigeon Creek	2	S	C	356	32	7	C	4	0.83	Default value	242

Site ID No. 1: K24, Wuhu, China.

Ref.: Yang et al. (2015a): Field behavior of driven Pre-stressed High-strength Concrete piles in sandy soils, Journal of Geotechnical and Geoenvironmental Engineering, ASCE, DOI: 10.1061/(ASCE)GT.1943-5606.0001303.

Cone penetrometer data

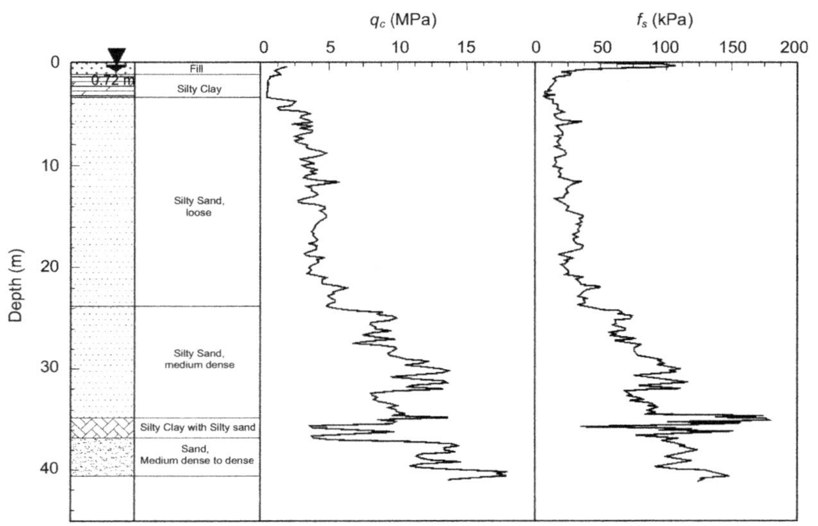

Detail	Description
Site name and location	Wuhu Second Bridge over the Yangtze River Test site: K24
Soil type (s)	Silty sand and fine sand
Water table depth (m)	0.72
Pile type (s)	Two circular PHC pile and one square pile
Type of cone penetrometer testing	Electric CPT, no pore pressure
Number of pile load tests	3
Comments	Three piles were driven in two phases: K24-1 in the first phase and the other two in the second phase. Interface friction angle estimated with PSD, and soil unit weight applies default value.

Pile ID: Wuhu K24-1

Load–displacement data

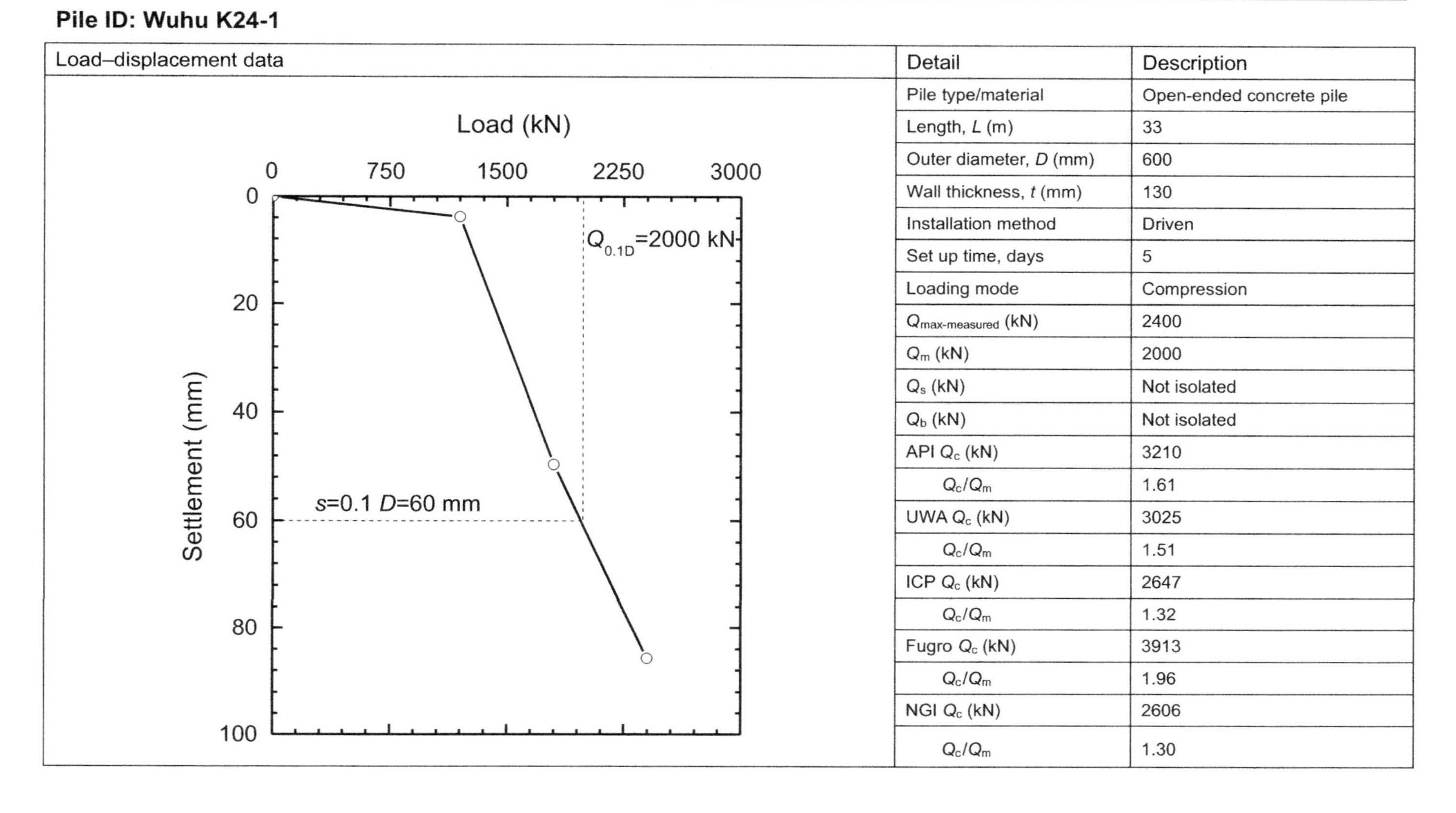

Detail	Description
Pile type/material	Open-ended concrete pile
Length, L (m)	33
Outer diameter, D (mm)	600
Wall thickness, t (mm)	130
Installation method	Driven
Set up time, days	5
Loading mode	Compression
$Q_{max\text{-}measured}$ (kN)	2400
Q_m (kN)	2000
Q_s (kN)	Not isolated
Q_b (kN)	Not isolated
API Q_c (kN)	3210
Q_c/Q_m	1.61
UWA Q_c (kN)	3025
Q_c/Q_m	1.51
ICP Q_c (kN)	2647
Q_c/Q_m	1.32
Fugro Q_c (kN)	3913
Q_c/Q_m	1.96
NGI Q_c (kN)	2606
Q_c/Q_m	1.30

Pile ID: Wuhu K24-2

Load–displacement data

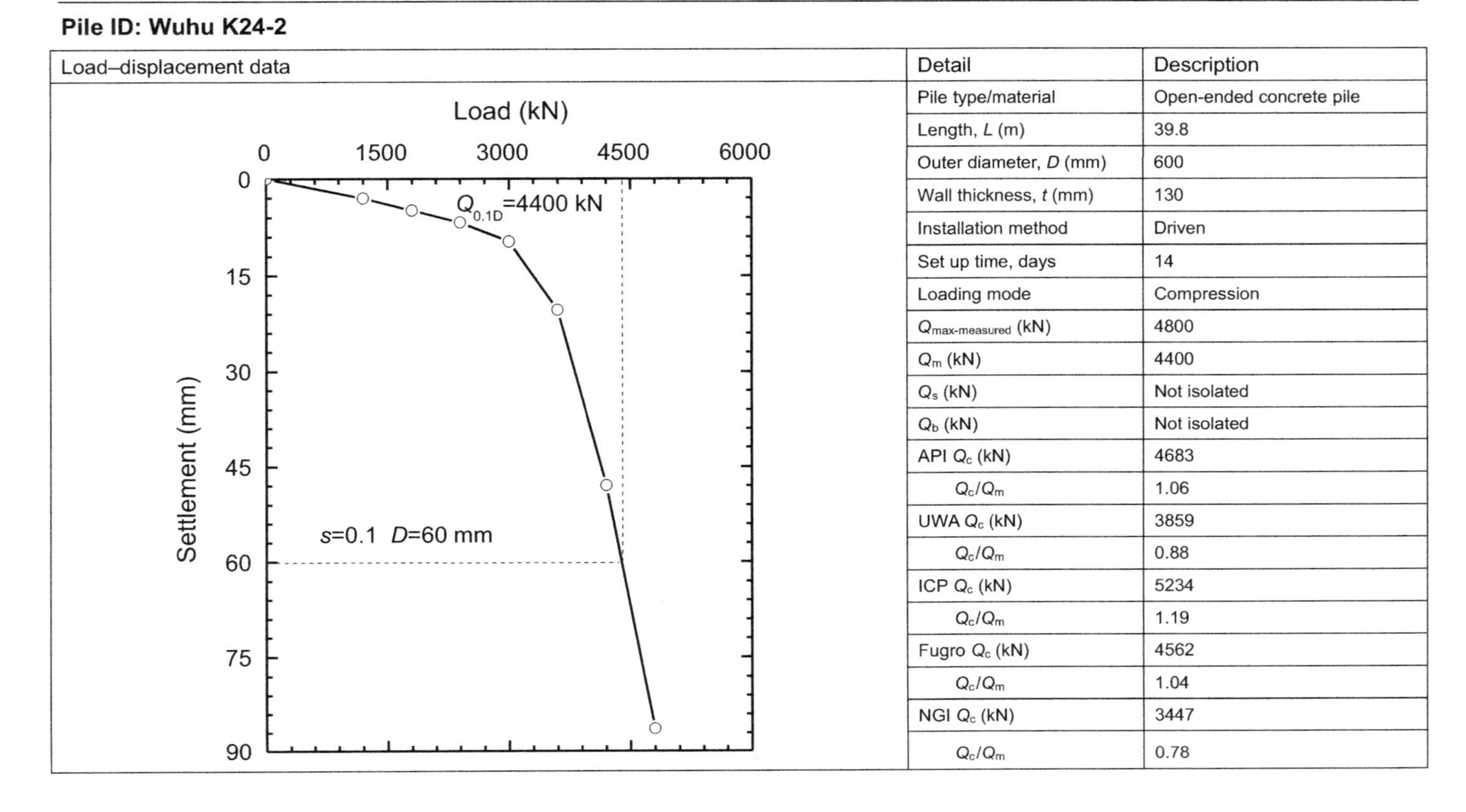

Detail	Description
Pile type/material	Open-ended concrete pile
Length, L (m)	39.8
Outer diameter, D (mm)	600
Wall thickness, t (mm)	130
Installation method	Driven
Set up time, days	14
Loading mode	Compression
$Q_{max\text{-}measured}$ (kN)	4800
Q_m (kN)	4400
Q_s (kN)	Not isolated
Q_b (kN)	Not isolated
API Q_c (kN)	4683
Q_c/Q_m	1.06
UWA Q_c (kN)	3859
Q_c/Q_m	0.88
ICP Q_c (kN)	5234
Q_c/Q_m	1.19
Fugro Q_c (kN)	4562
Q_c/Q_m	1.04
NGI Q_c (kN)	3447
Q_c/Q_m	0.78

Pile ID: Wuhu K24-3

Load–displacement data

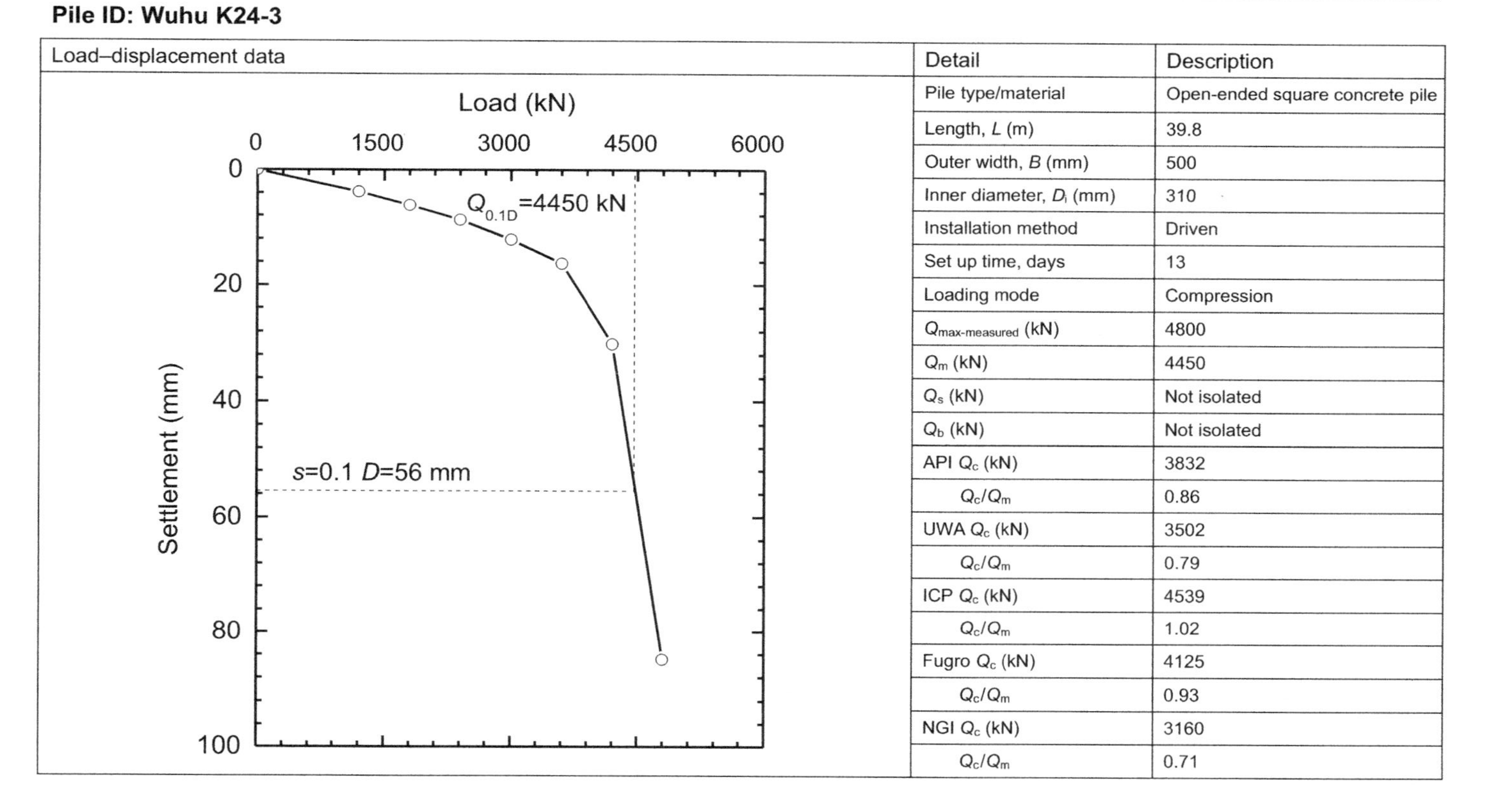

Detail	Description
Pile type/material	Open-ended square concrete pile
Length, L (m)	39.8
Outer width, B (mm)	500
Inner diameter, D_i (mm)	310
Installation method	Driven
Set up time, days	13
Loading mode	Compression
$Q_{max\text{-}measured}$ (kN)	4800
Q_m (kN)	4450
Q_s (kN)	Not isolated
Q_b (kN)	Not isolated
API Q_c (kN)	3832
Q_c/Q_m	0.86
UWA Q_c (kN)	3502
Q_c/Q_m	0.79
ICP Q_c (kN)	4539
Q_c/Q_m	1.02
Fugro Q_c (kN)	4125
Q_c/Q_m	0.93
NGI Q_c (kN)	3160
Q_c/Q_m	0.71

Site ID No. 2: K34, Wuhu, China.

Ref.: Yang et al. (2015a): Field behavior of driven Pre-stressed High-strength Concrete piles in sandy soils, Journal of Geotechnical and Geoenvironmental Engineering, ASCE, DOI: 10.1061/(ASCE)GT.1943-5606.0001303.

Cone penetrometer data

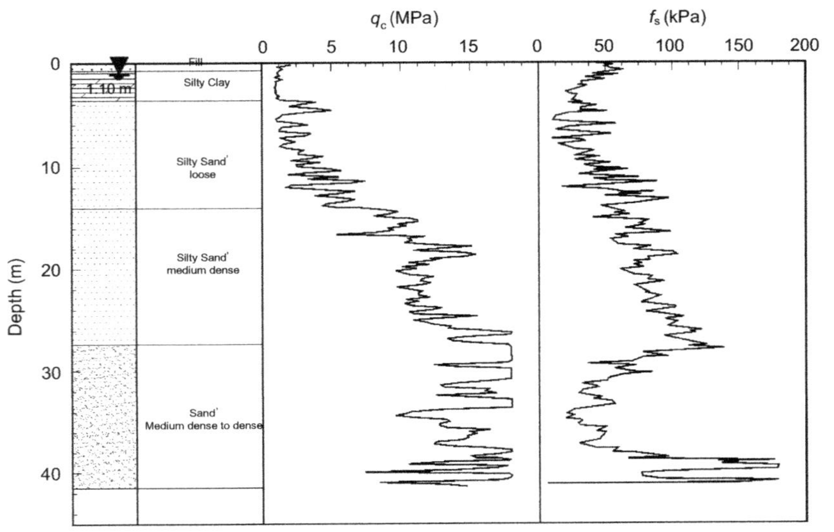

Detail	Description
Site name and location	Wuhu Second Bridge over the Yangtze River Test site: K34
Soil type (s)	Silty sand and fine sand
Water table depth (m)	1.1
Pile type (s)	PHC pile
Type of cone penetrometer testing	Electric CPT
Number of pile load tests	1
Comments	The q_c profile has three straight line segments of 1 to 2 m length where q_c exceeds the 18 MPa capacity of the CPT deployed. These sections are considered as q_c = 18 MPa in the analysis performed. Interface friction angle estimated with PSD, soil unit weight apply default value.

Pile ID: Wuhu K34-1

Load–displacement data

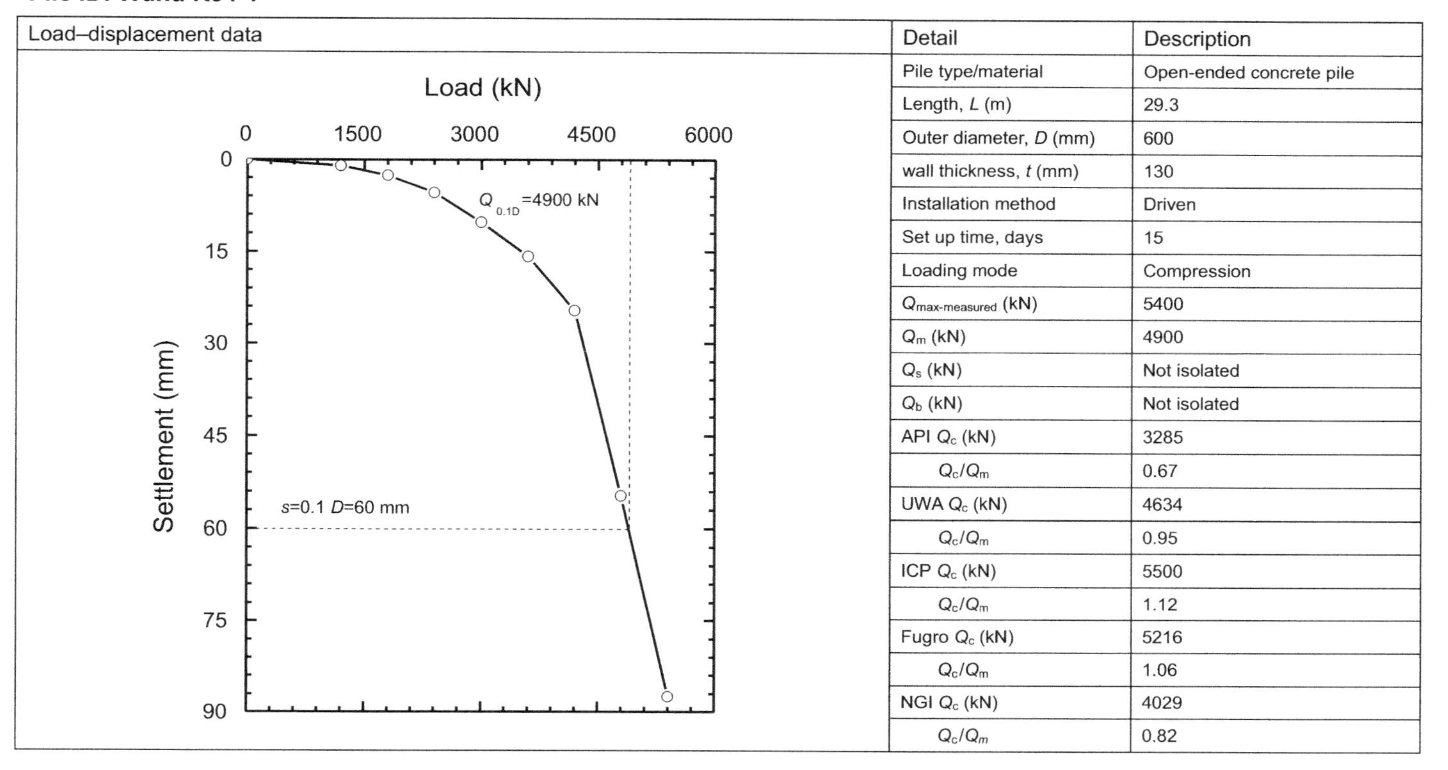

Detail	Description
Pile type/material	Open-ended concrete pile
Length, L (m)	29.3
Outer diameter, D (mm)	600
wall thickness, t (mm)	130
Installation method	Driven
Set up time, days	15
Loading mode	Compression
$Q_{max\text{-}measured}$ (kN)	5400
Q_m (kN)	4900
Q_s (kN)	Not isolated
Q_b (kN)	Not isolated
API Q_c (kN)	3285
Q_c/Q_m	0.67
UWA Q_c (kN)	4634
Q_c/Q_m	0.95
ICP Q_c (kN)	5500
Q_c/Q_m	1.12
Fugro Q_c (kN)	5216
Q_c/Q_m	1.06
NGI Q_c (kN)	4029
Q_c/Q_m	0.82

Site ID No. 3: K27, Wuhu, China.

Ref.: Yang et al. (2015a): Field behavior of driven Pre-stressed High-strength Concrete piles in sandy soils, Journal of Geotechnical and Geoenvironmental Engineering, ASCE, DOI: 10.1061/(ASCE)GT.1943-5606.0001303.

Cone penetrometer data

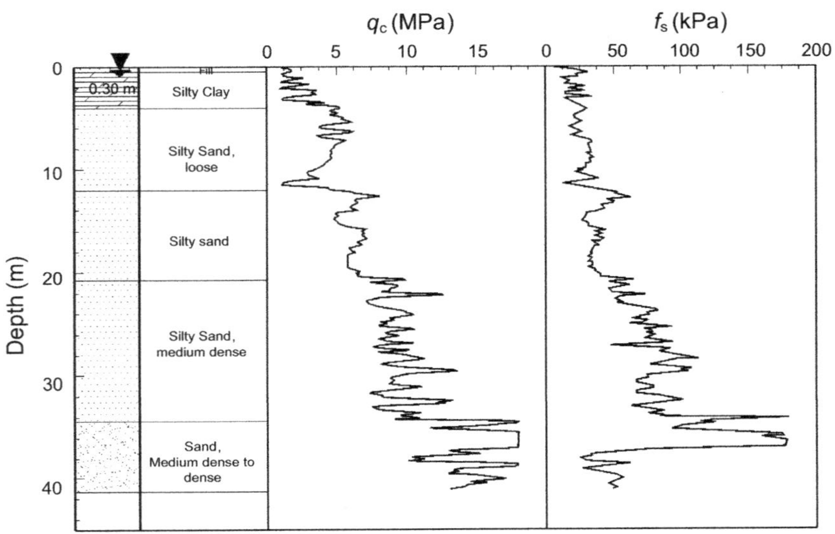

q_c (MPa)
0
5
10
15
f_s (kPa)
0
50
100
150
200
Depth (m)
0
10
20
30
40
0.30 m
Fill
Silty Clay
Silty Sand, loose
Silty sand
Silty Sand, medium dense
Sand, Medium dense to dense

Detail	Description
Site name and location	Wuhu Second Bridge over the Yangtze River Test site: K27
Soil type (s)	Silty sand and fine sand
Water table depth (m)	0.3
Pile type (s)	Pre-cast Hollow Concrete (PHC)
Type of cone penetrometer testing	Electric CPT
Number of pile load tests	1, with 3 others at other locations
Comments	The q_c profile has one straight line segments of 1 to 2 m length where q_c exceeds the 18 MPa capacity of the CPT deployed. These sections are considered as q_c = 18 MPa in the analysis performed. Interface friction angle estimated with PSD, and soil unit weight applies default value.

Pile ID: Wuhu K27-1

Load–displacement data

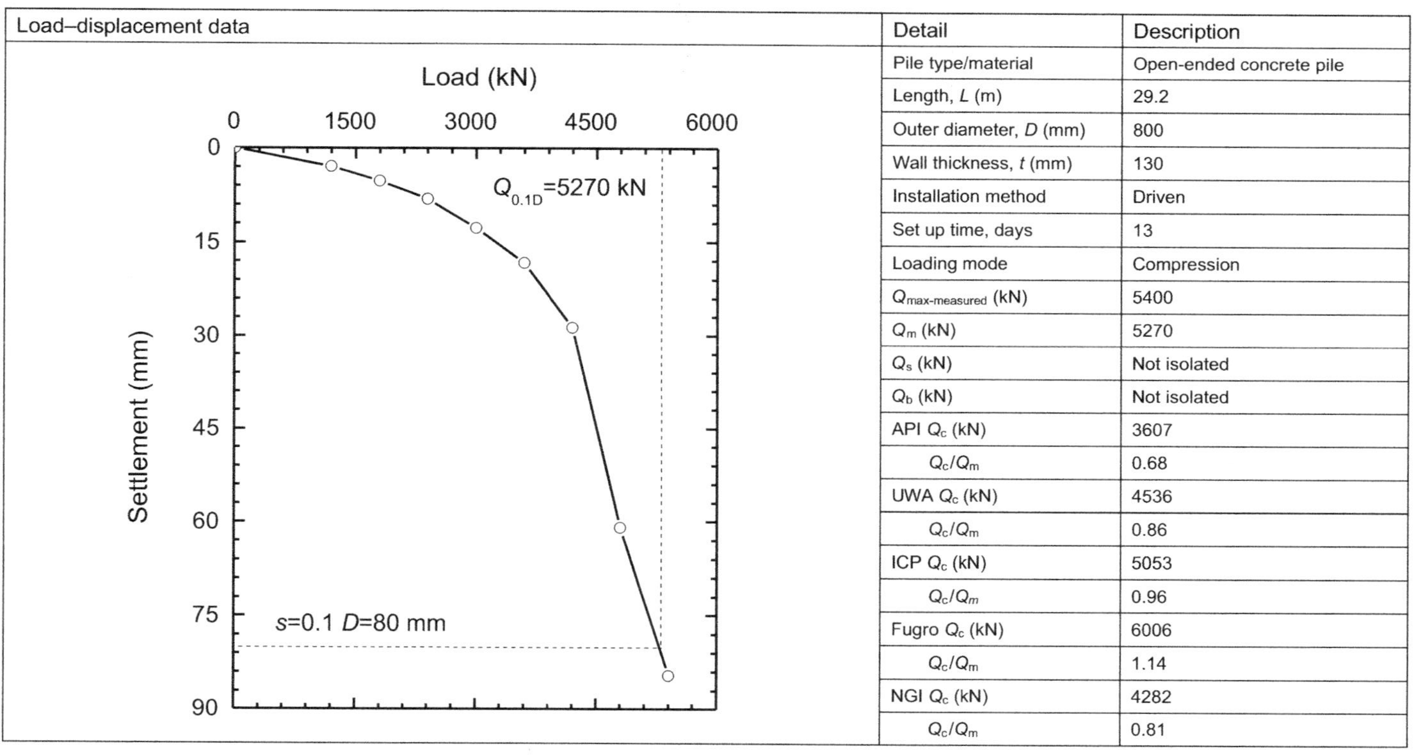

Detail	Description
Pile type/material	Open-ended concrete pile
Length, L (m)	29.2
Outer diameter, D (mm)	800
Wall thickness, t (mm)	130
Installation method	Driven
Set up time, days	13
Loading mode	Compression
$Q_{max\text{-}measured}$ (kN)	5400
Q_m (kN)	5270
Q_s (kN)	Not isolated
Q_b (kN)	Not isolated
API Q_c (kN)	3607
Q_c/Q_m	0.68
UWA Q_c (kN)	4536
Q_c/Q_m	0.86
ICP Q_c (kN)	5053
Q_c/Q_m	0.96
Fugro Q_c (kN)	6006
Q_c/Q_m	1.14
NGI Q_c (kN)	4282
Q_c/Q_m	0.81

Site ID No. 4: Rio de Janeiro, Brazil.

Ref.: Tsuha (2012): Companhia siderúrgica do atlântico, Rio de Janeiro. Private communication.

Cone penetrometer data

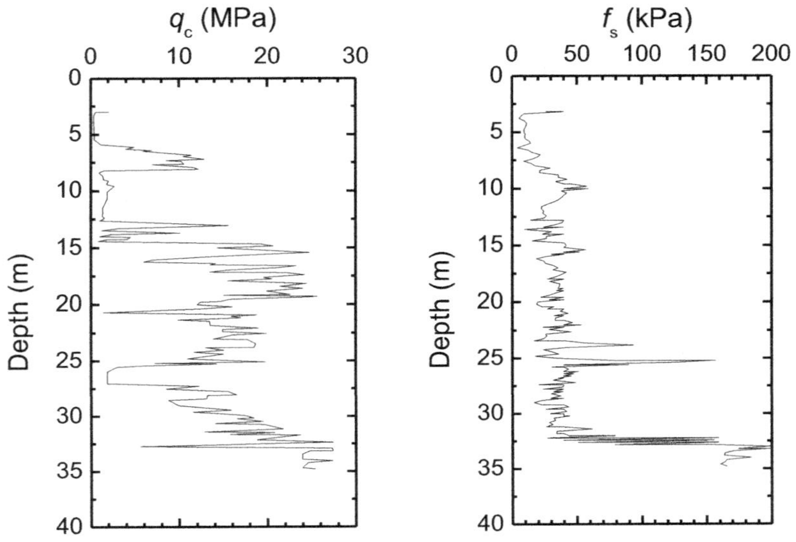

Detail	Description
Site name and location	Rio de Janeiro
Soil type (s)	Clay layer overlying silty and fine sand
Water table depth (m)	2.7
Pile type (s)	Square concrete pile
Type of cone penetrometer testing	CPTU with F_r and pore pressure
Number of pile load tests	2
Comments	The profile consists of about 12 m clay over thick sands. The two piles tested were un-instrumented. All piles were subjected to unloading and reloading cycles. Interface friction angle and soil unit weight all apply default value.

Pile ID: RJ PI-1

Load–displacement data

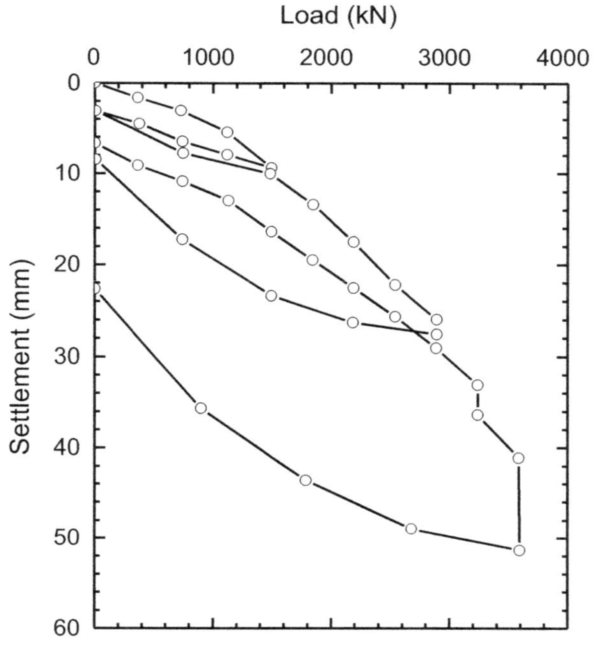

Detail	Description
Pile type/material	Square concrete pile
Length, L (m)	37.2
Outer width, B (mm)	500
Installation method	Driven
Set up time, days	64
Loading mode	Compression
$Q_{max\text{-}measured}$ (kN)	3590
Q_m (kN)	3590
Q_s (kN)	Not isolated
Q_b (kN)	Not isolated
API Q_c (kN)	3461
Q_c/Q_m	0.96
UWA Q_c (kN)	4336
Q_c/Q_m	1.21
ICP Q_c (kN)	3878
Q_c/Q_m	1.08
Fugro Q_c (kN)	4333
Q_c/Q_m	1.21
NGI Q_c (kN)	5535
Q_c/Q_m	1.54

Pile ID: RJ PI-2

Load–displacement data

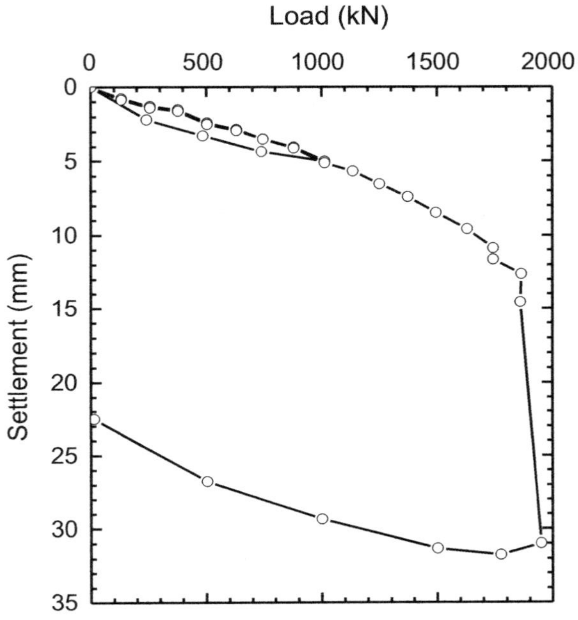

Detail	Description
Pile type/material	Square concrete pile
Length, L (m)	21.4
Outer width, B (mm)	500
Installation method	Driven
Set up time, days	72
Loading mode	Compression
$Q_{max\text{-}measured}$ (kN)	1950
Q_m (kN)	1950
Q_s (kN)	Not isolated
Q_b (kN)	Not isolated
API Q_c (kN)	1526
Q_c/Q_m	0.78
UWA Q_c (kN)	1700
Q_c/Q_m	0.87
ICP Q_c (kN)	1748
Q_c/Q_m	0.90
Fugro Q_c (kN)	2741
Q_c/Q_m	1.41
NGI Q_c (kN)	2889
Q_c/Q_m	1.48

Site ID No. 5: Rio de Janeiro, Brazil.

Ref.: Tsuha (2012): Companhia siderúrgica do atlântico, Rio de Janeiro. Private communication.

Cone penetrometer data

Detail	Description
Site name and location	Rio de Janeiro
Soil type (s)	Clay layer overlying silty and fine sand
Water table depth (m)	2.53
Pile type (s)	Square concrete pile
Type of cone penetrometer testing	CPTU with F_r and pore pressure
Number of pile load tests	2
Comments	The soil consists about 12 m deep clay layer, and the two piles at this site were instrumented. As Pile PI-3 was not loaded to failure, only the shaft capacity was used in the assessment and summaried in the next page with Q_c/Q_m referring to the shaft capacity. Interface friction angle and soil unit weight all apply default value.

Pile ID: RJ PI-3

Load–displacement data

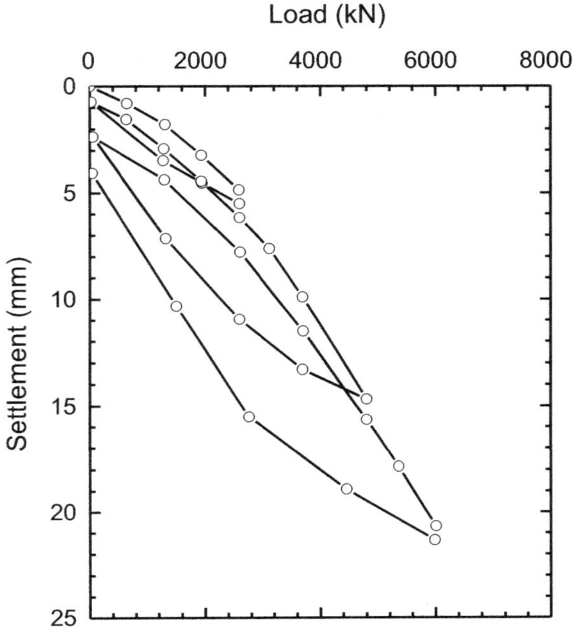

Load (kN)
0 2000 4000 6000 8000
Settlement (mm)
0 5 10 15 20 25

Detail	Description
Pile type/material	Square concrete pile
Length, L (m)	35.6
Outer width, B (mm)	700
Installation method	Driven
Set up time, days	89
Loading mode	Compression
$Q_{max\text{-}measured}$ (kN)	6010
Q_m (kN)	Not loaded to failure
Q_s (kN)	3680
Q_b (kN)	550 (not fully mobilized)
API Q_c (kN)	3611
Q_c/Q_m	0.98
UWA Q_c (kN)	3413
Q_c/Q_m	0.93
ICP Q_c (kN)	3419
Q_c/Q_m	0.93
Fugro Q_c (kN)	3197
Q_c/Q_m	0.87
NGI Q_c (kN)	4913
Q_c/Q_m	1.34

Pile ID: RJ PI-4

Load–displacement data

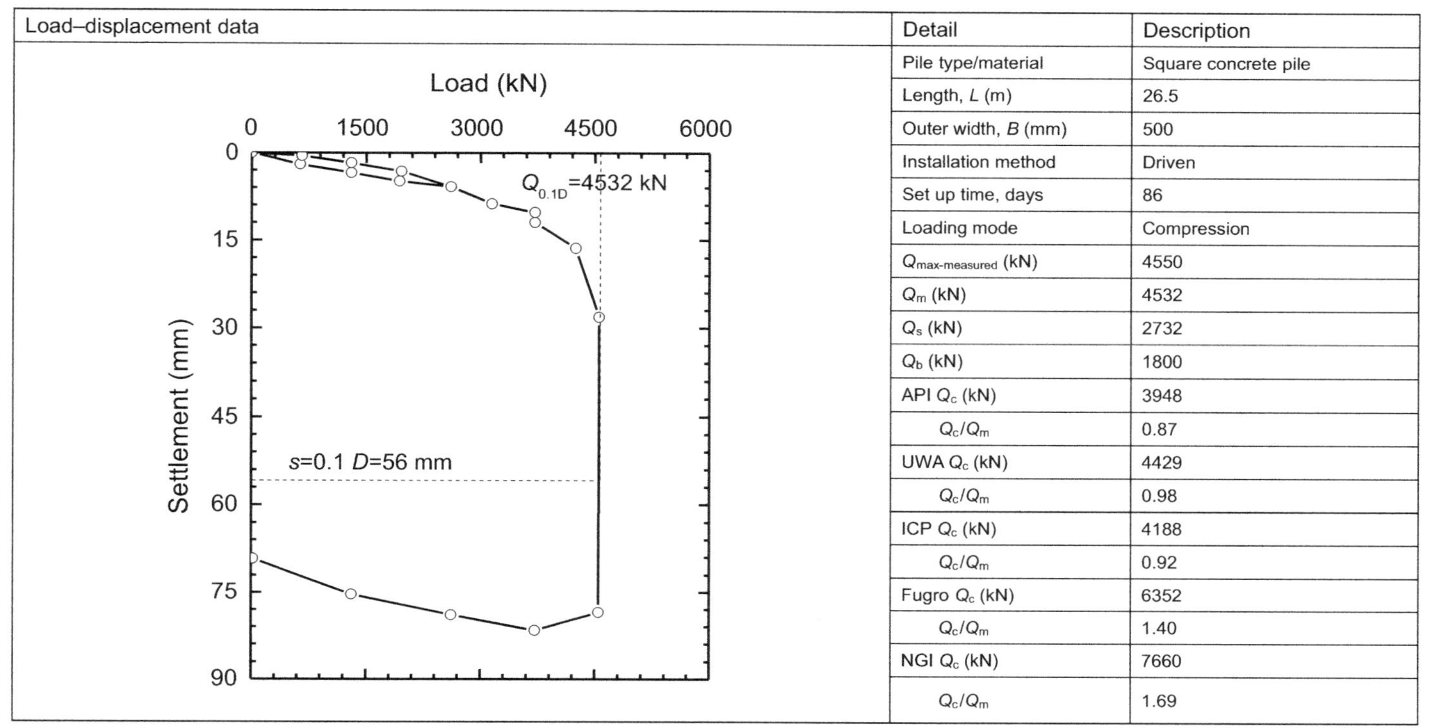

Detail	Description
Pile type/material	Square concrete pile
Length, L (m)	26.5
Outer width, B (mm)	500
Installation method	Driven
Set up time, days	86
Loading mode	Compression
$Q_{max\text{-}measured}$ (kN)	4550
Q_m (kN)	4532
Q_s (kN)	2732
Q_b (kN)	1800
API Q_c (kN)	3948
Q_c/Q_m	0.87
UWA Q_c (kN)	4429
Q_c/Q_m	0.98
ICP Q_c (kN)	4188
Q_c/Q_m	0.92
Fugro Q_c (kN)	6352
Q_c/Q_m	1.40
NGI Q_c (kN)	7660
Q_c/Q_m	1.69

Site ID No. 6: Blessington Dublin, Ireland.

Ref.: Gavin et al. (2013): The effect of ageing on the axial capacity of piles in sand. Proceedings of the ICE-Geotechnical Engineering, 166(2), 122–130.

Cone penetrometer data

Detail	Description
Site name and location	Blessington Dublin
Soil type (s)	Fine sand
Water table depth (m)	13
Pile type (s)	Open-ended steel piles
Type of cone penetrometer testing	CPT with no pore pressure
Number of pile load tests	3
Comments	Average q_c values used from the original CPT profiles. Very high interface friction angle measured with ring shear test (δ_f=36°), and unit weight applies the default value. IFR was also measured.

Pile ID: BD S2

Load–displacement data

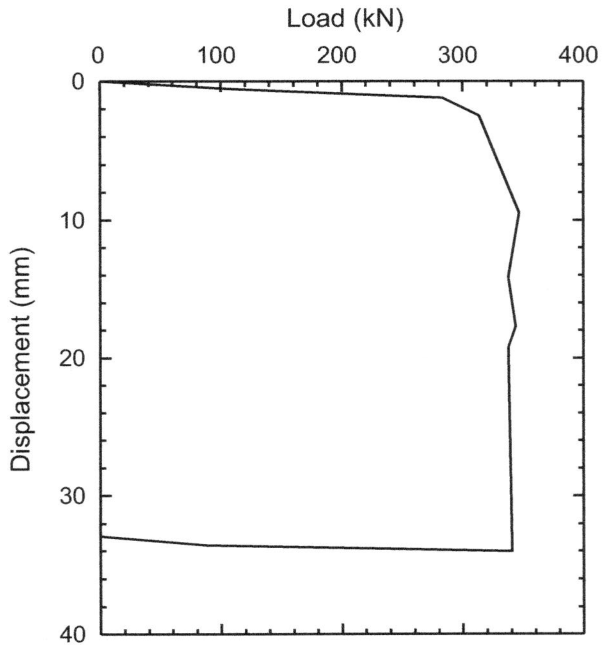

Detail	Description
Pile type/material	Open-ended steel piles
Length, L (m)	7
Outer diameter, D (mm)	340
Wall thickness, t (mm)	14
Installation method	Driven
Set up time, days	2
Loading mode	Tension
$Q_{max\text{-}measured}$ (kN)	344
Q_m (kN)	344
Q_s (kN)	344
Q_b (kN)	—
API Q_c (kN)	247
Q_c/Q_m	0.72
UWA Q_c (kN)	620
Q_c/Q_m	1.80
ICP Q_c (kN)	529
Q_c/Q_m	1.54
Fugro Q_c (kN)	644
Q_c/Q_m	1.87
NGI Q_c (kN)	459
Q_c/Q_m	1.33

Site ID No. 6

Pile ID: BD S3

Load–displacement data

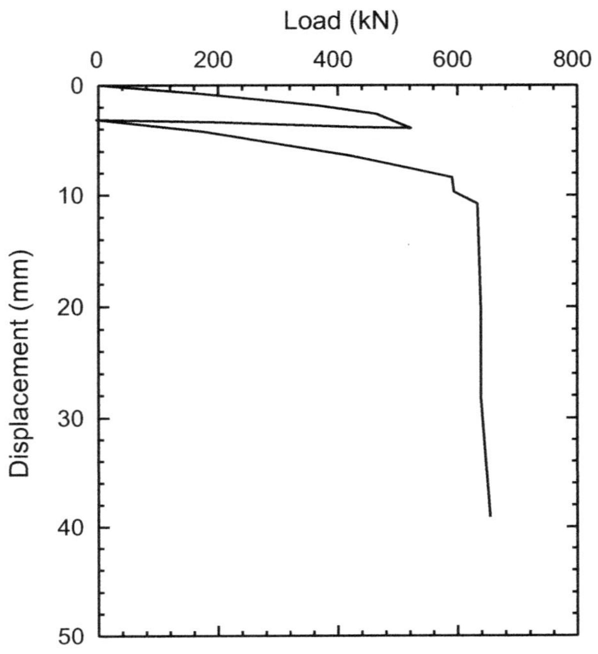

Load (kN)
0 200 400 600 800
Displacement (mm)
0 10 20 30 40 50

Detail	Description
Pile type/material	Open-ended steel piles
Length, L (m)	7
Outer diameter, D (mm)	340
Wall thickness, t (mm)	14
Installation method	Driven
Set up time, days	13
Loading mode	Tension
$Q_{max\text{-}measured}$ (kN)	665
Q_m (kN)	665
Q_s (kN)	665
Q_b (kN)	—
API Q_c (kN)	251
Q_c/Q_m	0.38
UWA Q_c (kN)	673
Q_c/Q_m	1.01
ICP Q_c (kN)	573
Q_c/Q_m	0.85
Fugro Q_c (kN)	702
Q_c/Q_m	1.06
NGI Q_c (kN)	491
Q_c/Q_m	0.74

Pile ID: BD S5

Load–displacement data

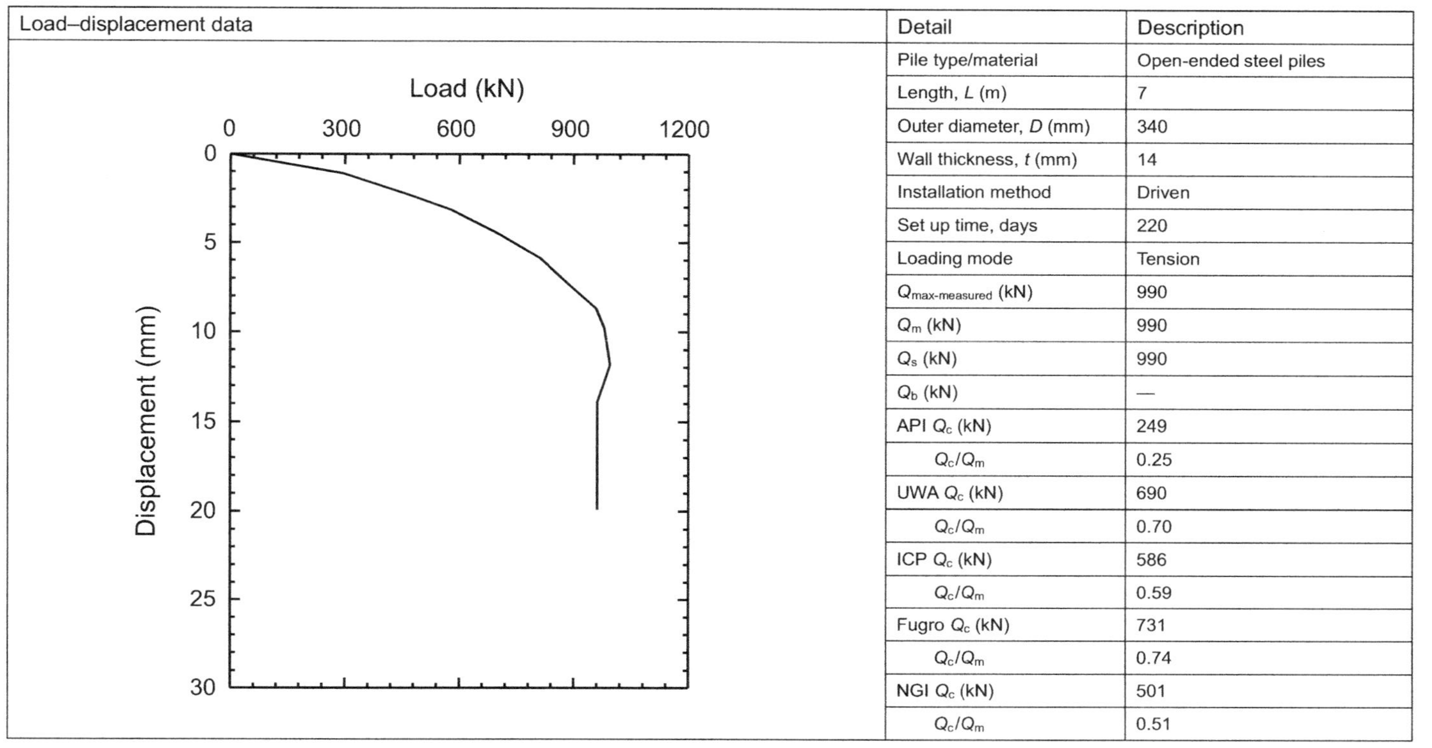

Detail	Description
Pile type/material	Open-ended steel piles
Length, L (m)	7
Outer diameter, D (mm)	340
Wall thickness, t (mm)	14
Installation method	Driven
Set up time, days	220
Loading mode	Tension
$Q_{max\text{-}measured}$ (kN)	990
Q_m (kN)	990
Q_s (kN)	990
Q_b (kN)	—
API Q_c (kN)	249
Q_c/Q_m	0.25
UWA Q_c (kN)	690
Q_c/Q_m	0.70
ICP Q_c (kN)	586
Q_c/Q_m	0.59
Fugro Q_c (kN)	731
Q_c/Q_m	0.74
NGI Q_c (kN)	501
Q_c/Q_m	0.51

Site ID No. 7: Horstwalde, Germany.

Ref.: Rücker et al. (2013): Großversuche an Rammpfählen zur Ermittlung der Tragfähigkeit unter zyklischer Belastung und Standzeit. Geotechnik, 36(2), 77–89.

Cone penetrometer data

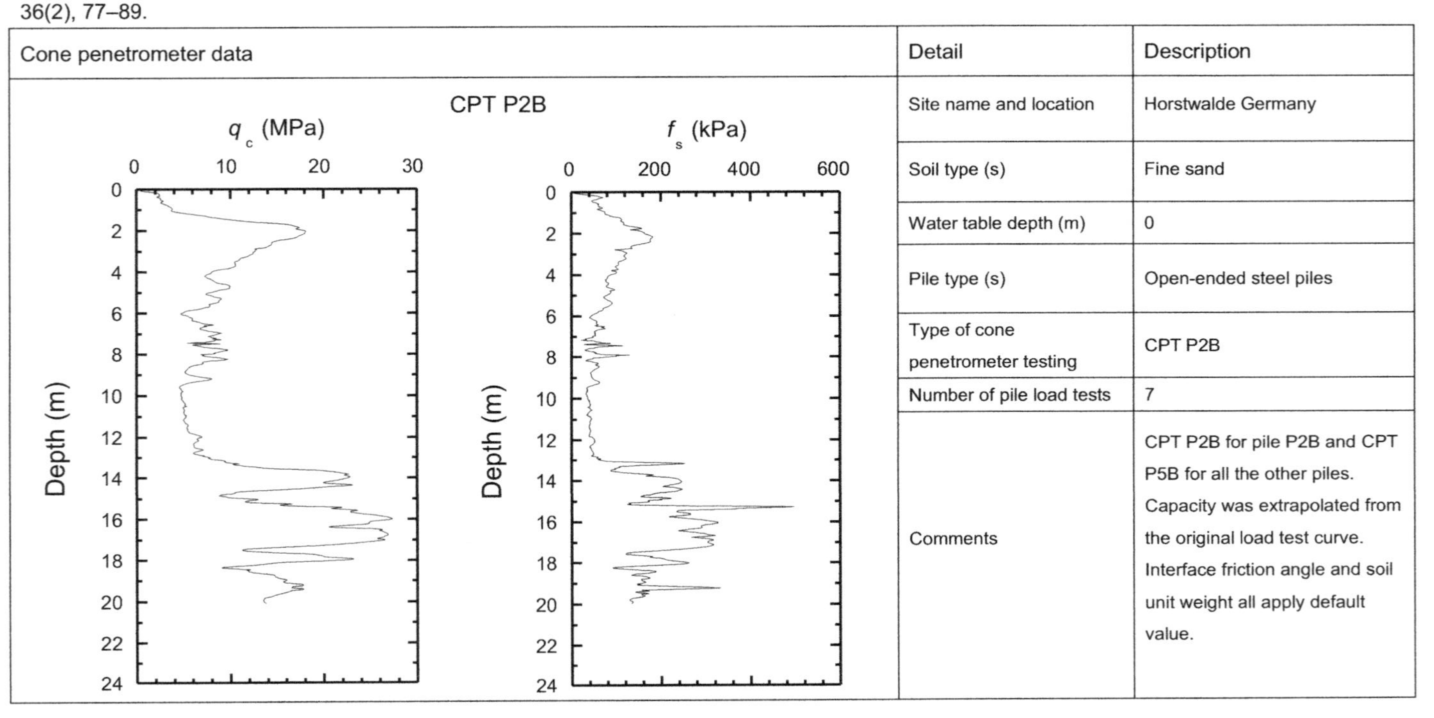

Detail	Description
Site name and location	Horstwalde Germany
Soil type (s)	Fine sand
Water table depth (m)	0
Pile type (s)	Open-ended steel piles
Type of cone penetrometer testing	CPT P2B
Number of pile load tests	7
Comments	CPT P2B for pile P2B and CPT P5B for all the other piles. Capacity was extrapolated from the original load test curve. Interface friction angle and soil unit weight all apply default value.

Site ID No. 7

Cone penetrometer data	Detail	Description
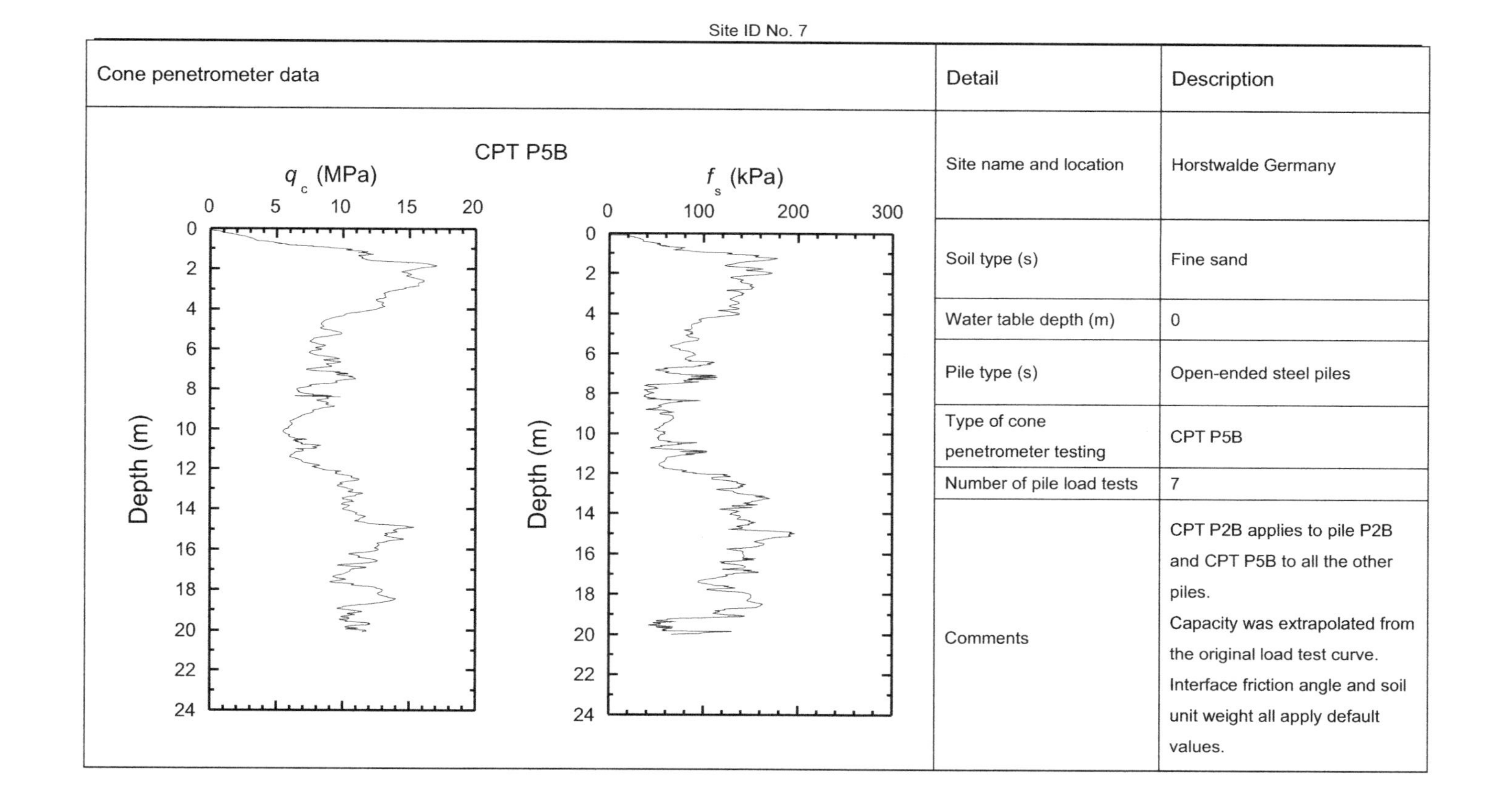	Site name and location	Horstwalde Germany
	Soil type (s)	Fine sand
	Water table depth (m)	0
	Pile type (s)	Open-ended steel piles
	Type of cone penetrometer testing	CPT P5B
	Number of pile load tests	7
	Comments	CPT P2B applies to pile P2B and CPT P5B to all the other piles. Capacity was extrapolated from the original load test curve. Interface friction angle and soil unit weight all apply default values.

Pile ID: HG P2B

Load–displacement data

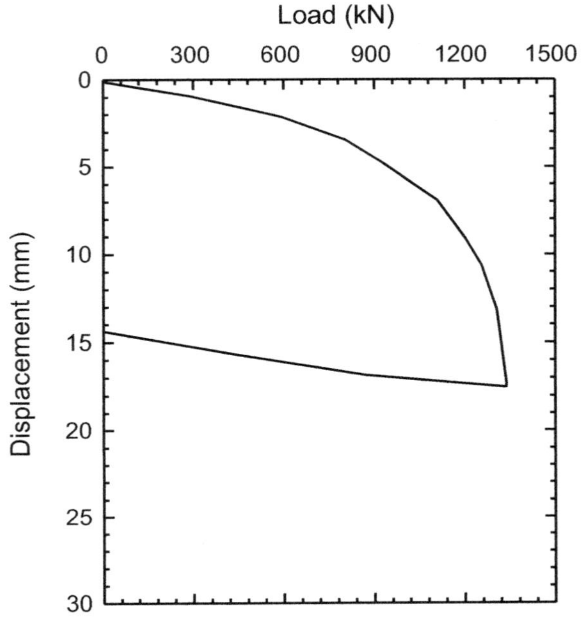

Detail	Description
Pile type/material	Open-ended steel piles
Length, L (m)	17.61
Outer diameter, D (mm)	711
Wall thickness, t (mm)	12.5
Installation method	Driven
Set up time, days	43
Loading mode	Tension
$Q_{max\text{-}measured}$ (kN)	1400
Q_m (kN)	1400
Q_s (kN)	1400
Q_b (kN)	—
API Q_c (kN)	1268
Q_c/Q_m	0.90
UWA Q_c (kN)	1625
Q_c/Q_m	1.15
ICP Q_c (kN)	1323
Q_c/Q_m	0.94
Fugro Q_c (kN)	2661
Q_c/Q_m	1.90
NGI Q_c (kN)	1560
Q_c/Q_m	1.11

Pile ID: HG P2D

Load–displacement data

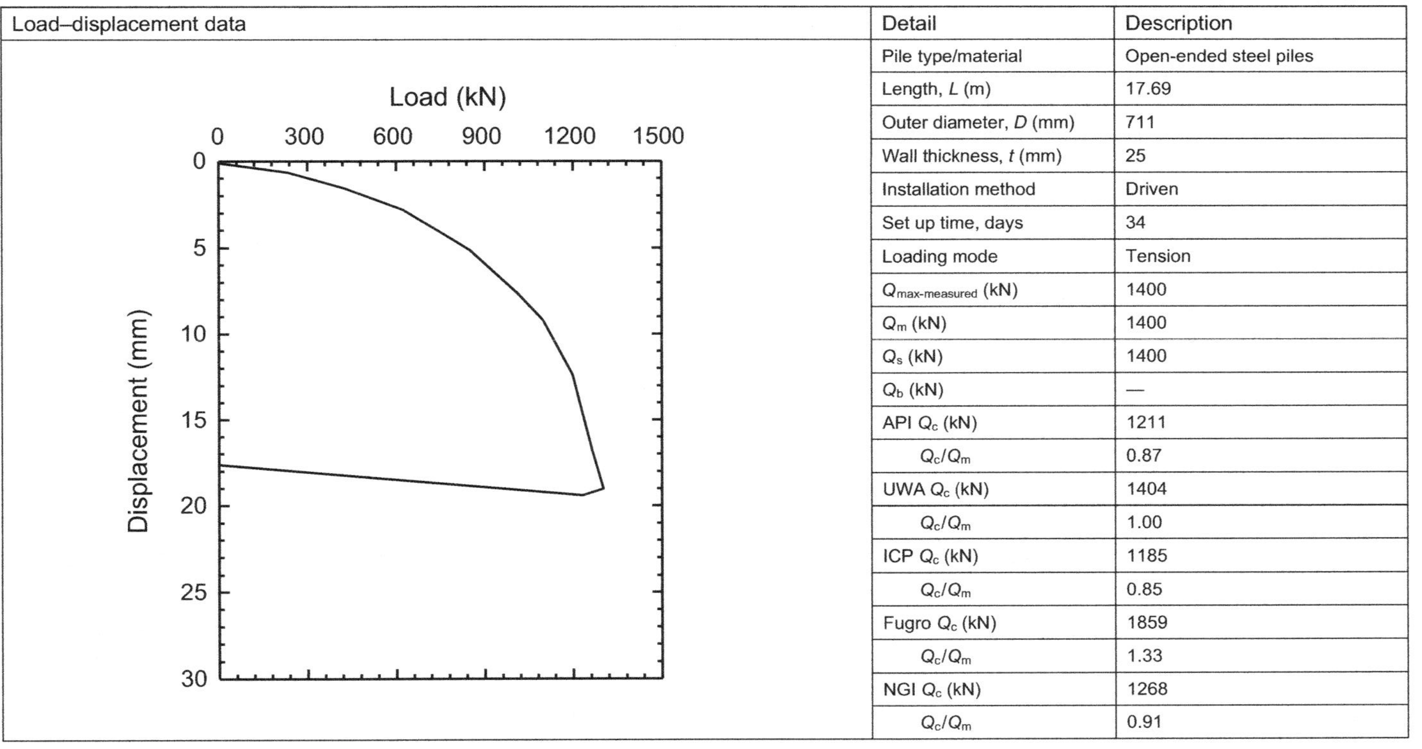

Detail	Description
Pile type/material	Open-ended steel piles
Length, L (m)	17.69
Outer diameter, D (mm)	711
Wall thickness, t (mm)	25
Installation method	Driven
Set up time, days	34
Loading mode	Tension
$Q_{max\text{-}measured}$ (kN)	1400
Q_m (kN)	1400
Q_s (kN)	1400
Q_b (kN)	—
API Q_c (kN)	1211
Q_c/Q_m	0.87
UWA Q_c (kN)	1404
Q_c/Q_m	1.00
ICP Q_c (kN)	1185
Q_c/Q_m	0.85
Fugro Q_c (kN)	1859
Q_c/Q_m	1.33
NGI Q_c (kN)	1268
Q_c/Q_m	0.91

Site ID No. 7

Pile ID: HG P5B

Load–displacement data

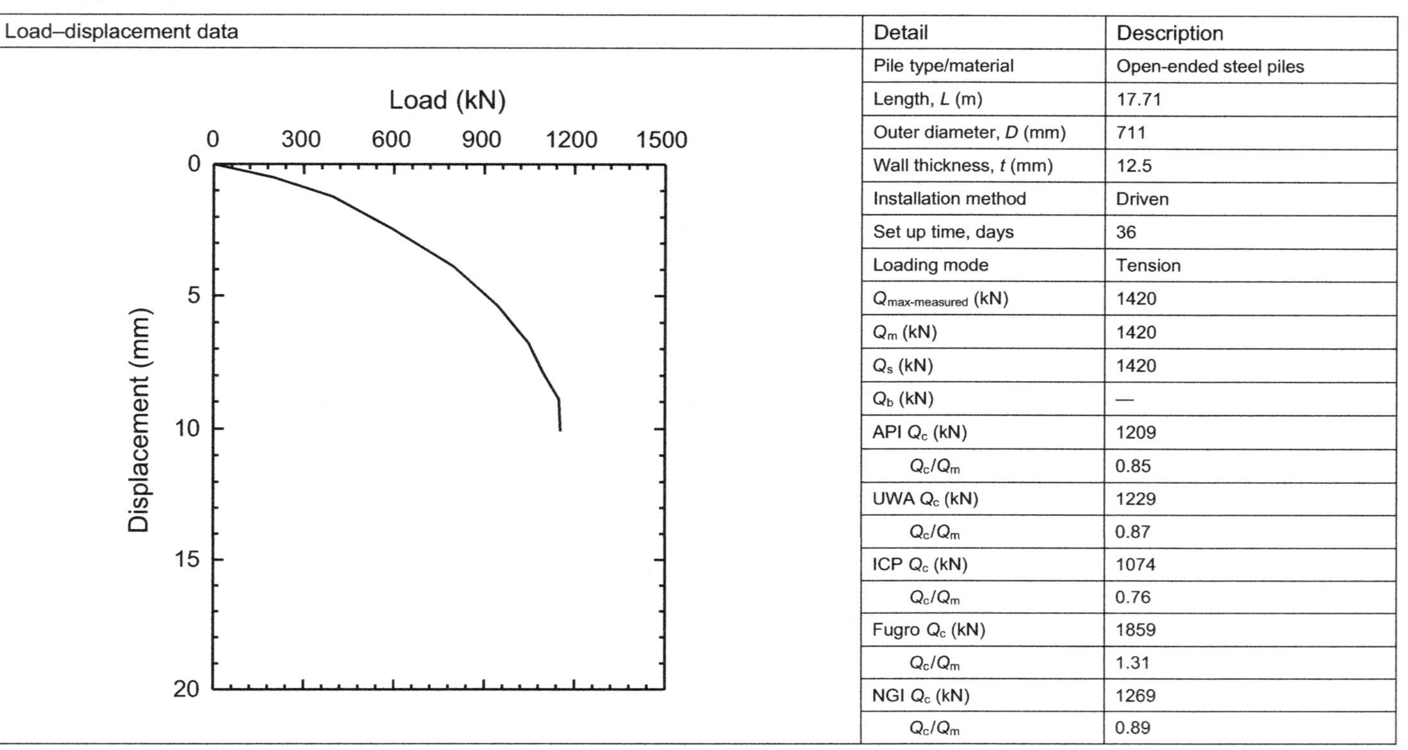

Detail	Description
Pile type/material	Open-ended steel piles
Length, L (m)	17.71
Outer diameter, D (mm)	711
Wall thickness, t (mm)	12.5
Installation method	Driven
Set up time, days	36
Loading mode	Tension
$Q_{max\text{-}measured}$ (kN)	1420
Q_m (kN)	1420
Q_s (kN)	1420
Q_b (kN)	—
API Q_c (kN)	1209
Q_c/Q_m	0.85
UWA Q_c (kN)	1229
Q_c/Q_m	0.87
ICP Q_c (kN)	1074
Q_c/Q_m	0.76
Fugro Q_c (kN)	1859
Q_c/Q_m	1.31
NGI Q_c (kN)	1269
Q_c/Q_m	0.89

Pile ID: HG P5D

Load–displacement data

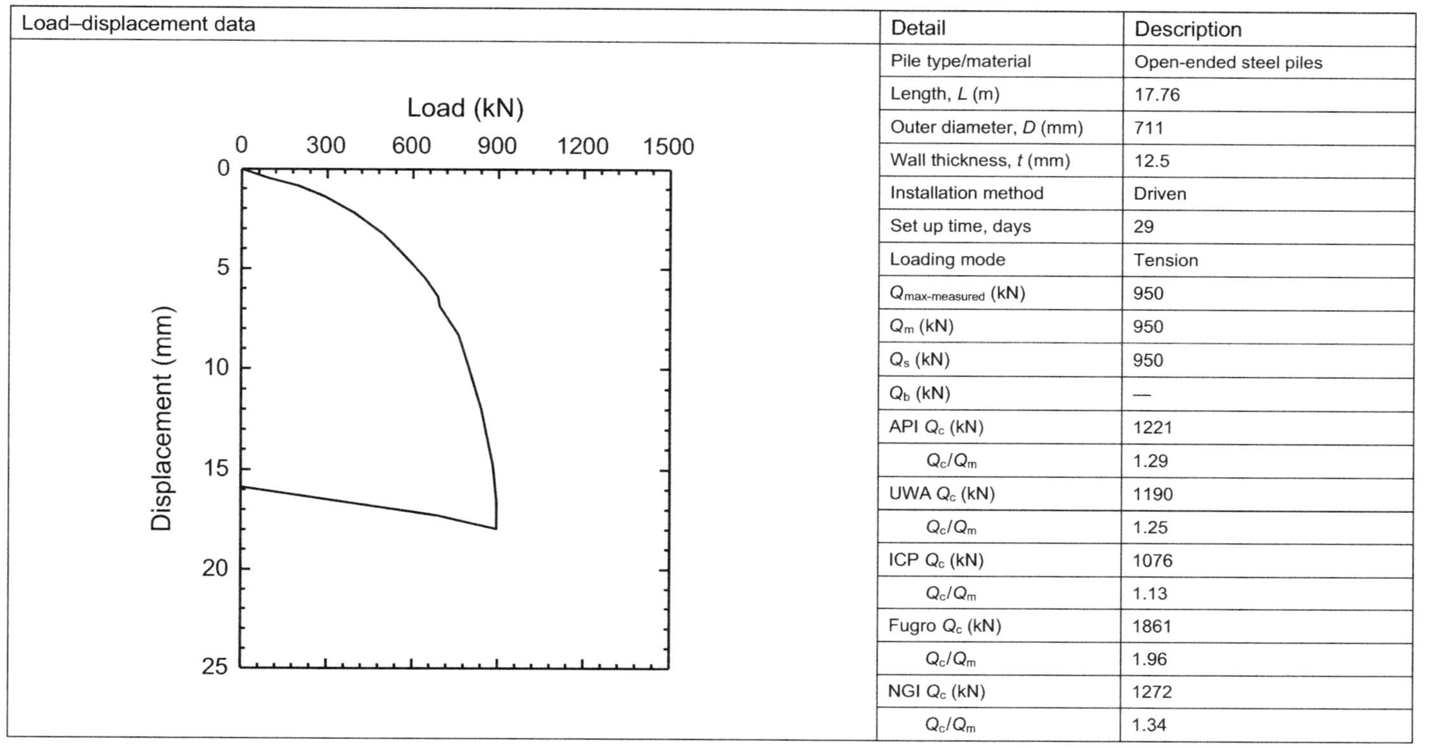

Detail	Description
Pile type/material	Open-ended steel piles
Length, L (m)	17.76
Outer diameter, D (mm)	711
Wall thickness, t (mm)	12.5
Installation method	Driven
Set up time, days	29
Loading mode	Tension
$Q_{max\text{-}measured}$ (kN)	950
Q_m (kN)	950
Q_s (kN)	950
Q_b (kN)	—
API Q_c (kN)	1221
Q_c/Q_m	1.29
UWA Q_c (kN)	1190
Q_c/Q_m	1.25
ICP Q_c (kN)	1076
Q_c/Q_m	1.13
Fugro Q_c (kN)	1861
Q_c/Q_m	1.96
NGI Q_c (kN)	1272
Q_c/Q_m	1.34

Pile ID: HG P4B

Load–displacement data

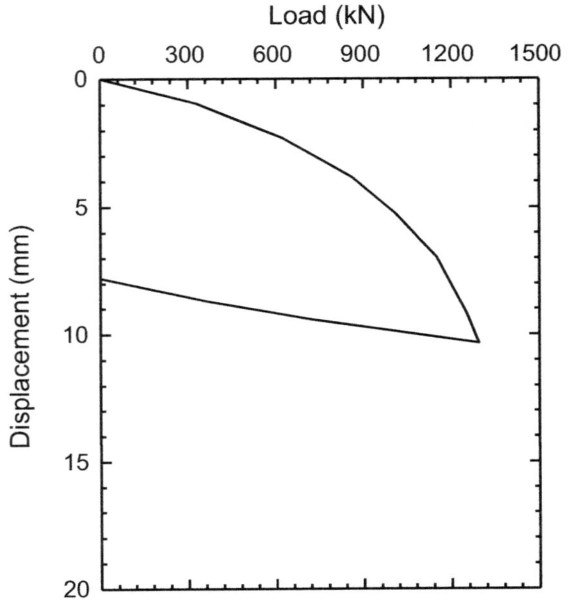

Load (kN)
0 300 600 900 1200 1500
Displacement (mm)
0 5 10 15 20

Detail	Description
Pile type/material	Open-ended steel piles
Length, L (m)	17.67
Outer diameter, D (mm)	711
Wall thickness, t (mm)	12.5
Installation method	Driven
Set up time, days	37
Loading mode	Tension
$Q_{max\text{-}measured}$ (kN)	1550
Q_m (kN)	1550
Q_s (kN)	1550
Q_b (kN)	—
API Q_c (kN)	1278
Q_c/Q_m	0.82
UWA Q_c (kN)	1572
Q_c/Q_m	1.01
ICP Q_c (kN)	1326
Q_c/Q_m	0.86
Fugro Q_c (kN)	2666
Q_c/Q_m	1.72
NGI Q_c (kN)	1568
Q_c/Q_m	1.01

Pile ID: HG P4D

Load–displacement data

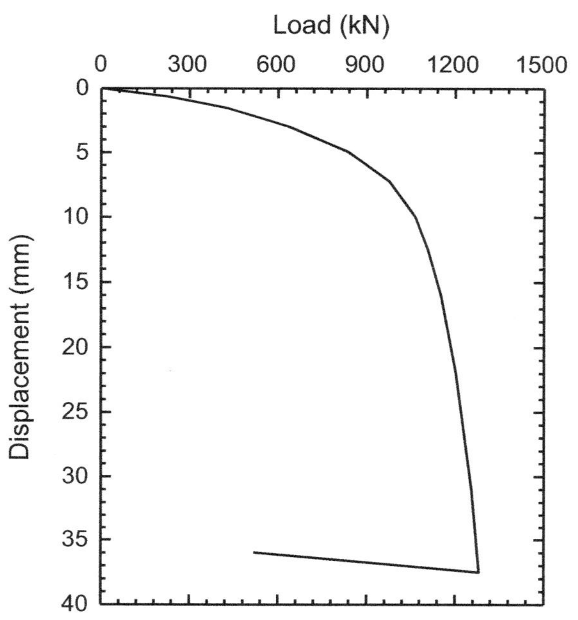

Detail	Description
Pile type/material	Open-ended steel piles
Length, L (m)	17.66
Outer diameter, D (mm)	711
Wall thickness, t (mm)	12.5
Installation method	Driven
Set up time, days	32
Loading mode	Tension
$Q_{max\text{-}measured}$ (kN)	1250
Q_m (kN)	1250
Q_s (kN)	1250
Q_b (kN)	—
API Q_c (kN)	1202
Q_c/Q_m	0.96
UWA Q_c (kN)	1213
Q_c/Q_m	0.97
ICP Q_c (kN)	1071
Q_c/Q_m	0.86
Fugro Q_c (kN)	1858
Q_c/Q_m	1.49
NGI Q_c (kN)	1267
Q_c/Q_m	1.01

Site ID No. 7

Pile ID: HG P3B

Load–displacement data

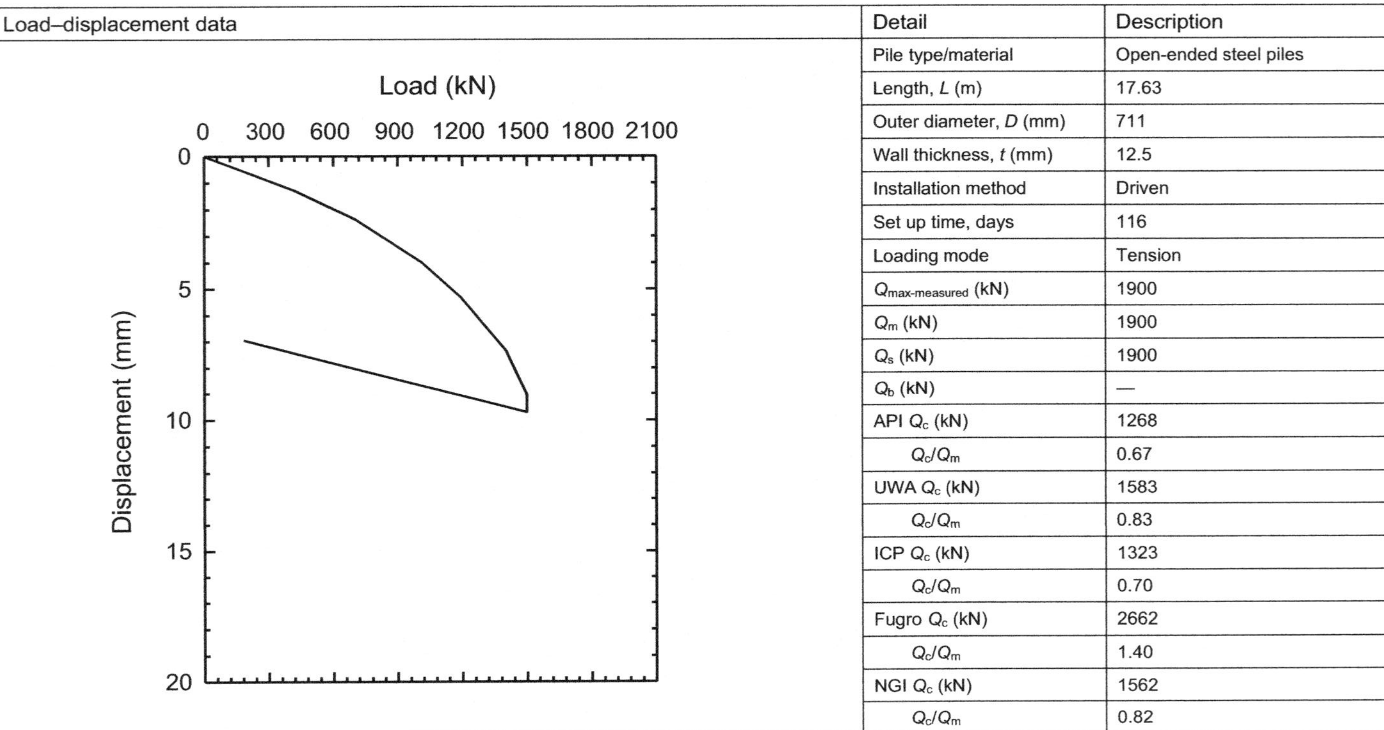

Detail	Description
Pile type/material	Open-ended steel piles
Length, L (m)	17.63
Outer diameter, D (mm)	711
Wall thickness, t (mm)	12.5
Installation method	Driven
Set up time, days	116
Loading mode	Tension
$Q_{max\text{-}measured}$ (kN)	1900
Q_m (kN)	1900
Q_s (kN)	1900
Q_b (kN)	—
API Q_c (kN)	1268
Q_c/Q_m	0.67
UWA Q_c (kN)	1583
Q_c/Q_m	0.83
ICP Q_c (kN)	1323
Q_c/Q_m	0.70
Fugro Q_c (kN)	2662
Q_c/Q_m	1.40
NGI Q_c (kN)	1562
Q_c/Q_m	0.82

Pile ID: HG P3D

Load–displacement data

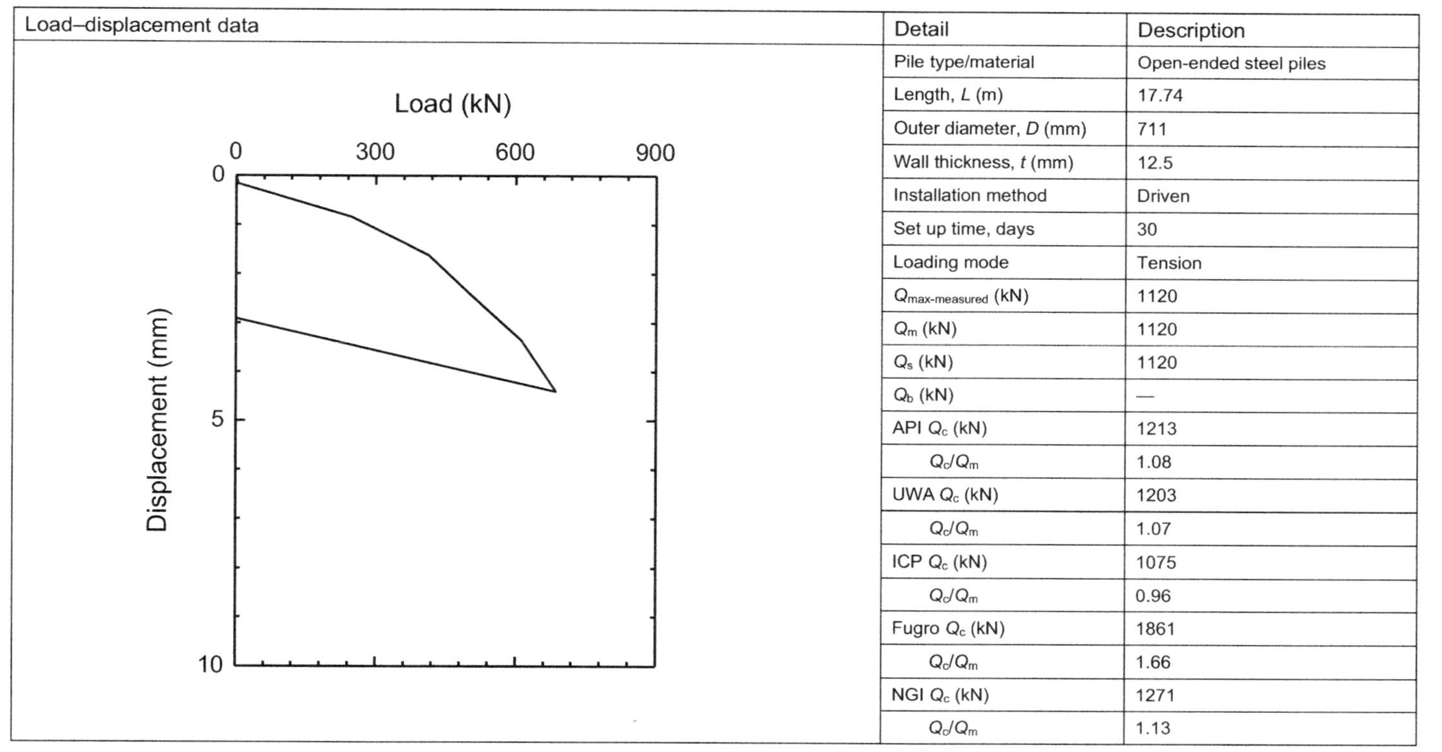

Detail	Description
Pile type/material	Open-ended steel piles
Length, L (m)	17.74
Outer diameter, D (mm)	711
Wall thickness, t (mm)	12.5
Installation method	Driven
Set up time, days	30
Loading mode	Tension
$Q_{max\text{-}measured}$ (kN)	1120
Q_m (kN)	1120
Q_s (kN)	1120
Q_b (kN)	—
API Q_c (kN)	1213
Q_c/Q_m	1.08
UWA Q_c (kN)	1203
Q_c/Q_m	1.07
ICP Q_c (kN)	1075
Q_c/Q_m	0.96
Fugro Q_c (kN)	1861
Q_c/Q_m	1.66
NGI Q_c (kN)	1271
Q_c/Q_m	1.13

Site ID No. 8: British Columbia, Canada.

Ref.: Naesgaard et al. (2012): Long piles in thick lacustrine and deltaic deposits. Two bridge foundation case histories. Geotechnical Special Publication, No. 227, ASCE, 404–421.

Cone penetrometer data

q_c (MPa)

Depth (m)

f_s (kPa)

Depth (m)

Detail	Description
Site name and location	W. R. Bennett Bridge, Okanagan Lake at Kelowna, BC, Canada
Soil type (s)	Loose to medium dense silts and sandy silt
Water table depth (m)	0
Pile type (s)	Closed-ended steel pile
Type of cone penetrometer testing	CPTU with F_r and pore pressure
Number of pile load tests	1
Comments	Total five piles with one static test accepted. The others are PDA tests. Interface friction angle and soil unit weight apply default value.

Pile ID: BC P1

Load–displacement data

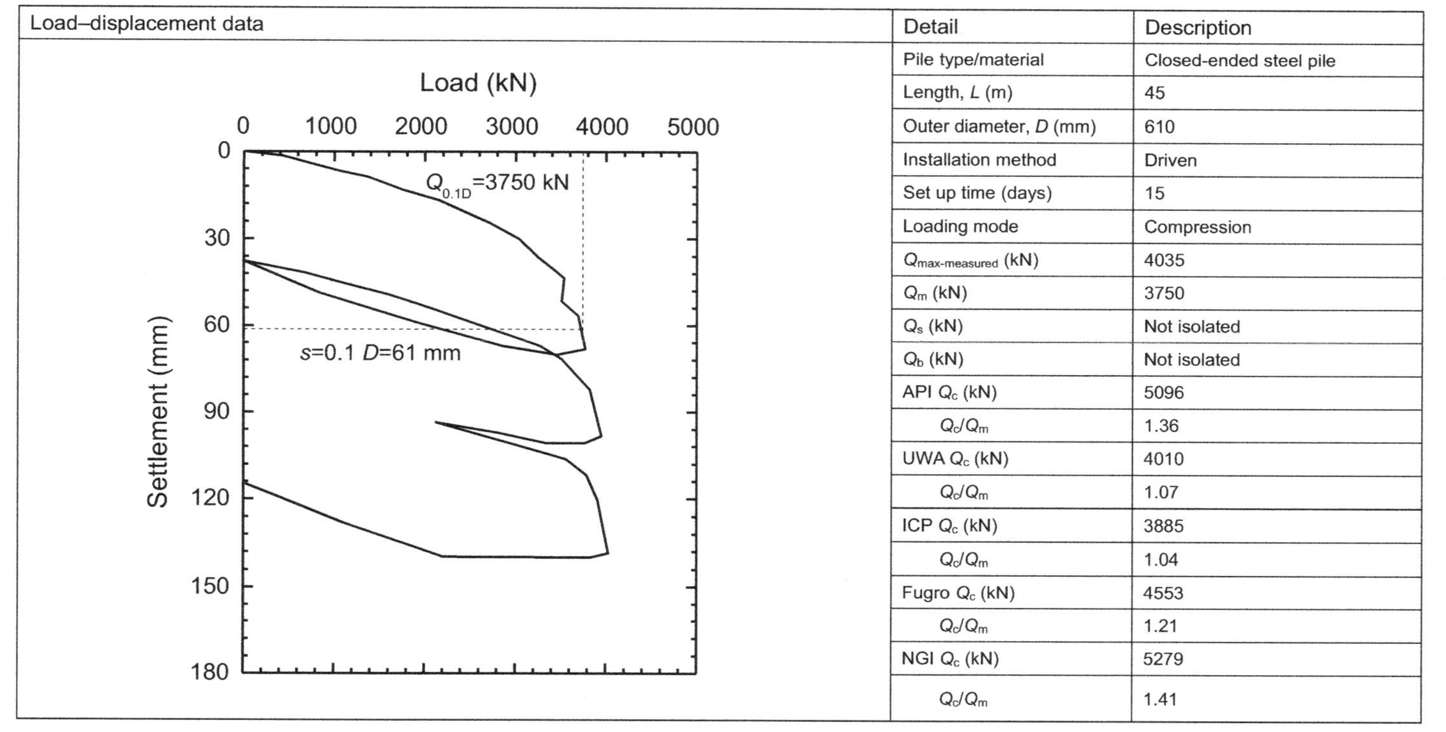

Detail	Description
Pile type/material	Closed-ended steel pile
Length, L (m)	45
Outer diameter, D (mm)	610
Installation method	Driven
Set up time (days)	15
Loading mode	Compression
$Q_{max\text{-}measured}$ (kN)	4035
Q_m (kN)	3750
Q_s (kN)	Not isolated
Q_b (kN)	Not isolated
API Q_c (kN)	5096
Q_c/Q_m	1.36
UWA Q_c (kN)	4010
Q_c/Q_m	1.07
ICP Q_c (kN)	3885
Q_c/Q_m	1.04
Fugro Q_c (kN)	4553
Q_c/Q_m	1.21
NGI Q_c (kN)	5279
Q_c/Q_m	1.41

Site ID No. 9: Hampton Virginia, USA.

Ref.: Pando et al. (2003): Axial and lateral load performance of two composite piles and one prestressed concrete pile. TRB 2003 Annual Meeting, 22p.

Cone penetrometer data

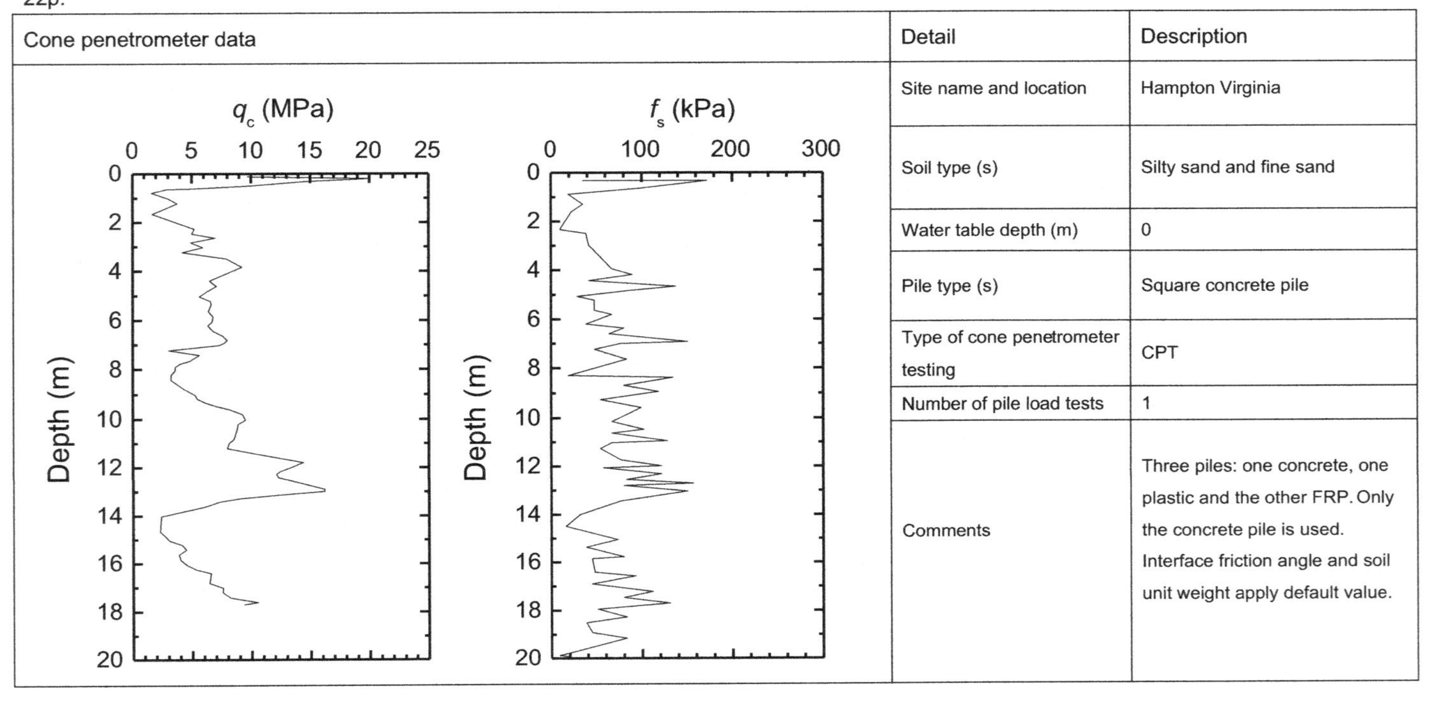

Detail	Description
Site name and location	Hampton Virginia
Soil type (s)	Silty sand and fine sand
Water table depth (m)	0
Pile type (s)	Square concrete pile
Type of cone penetrometer testing	CPT
Number of pile load tests	1
Comments	Three piles: one concrete, one plastic and the other FRP. Only the concrete pile is used. Interface friction angle and soil unit weight apply default value.

Pile ID: HRV P1

Load–displacement data

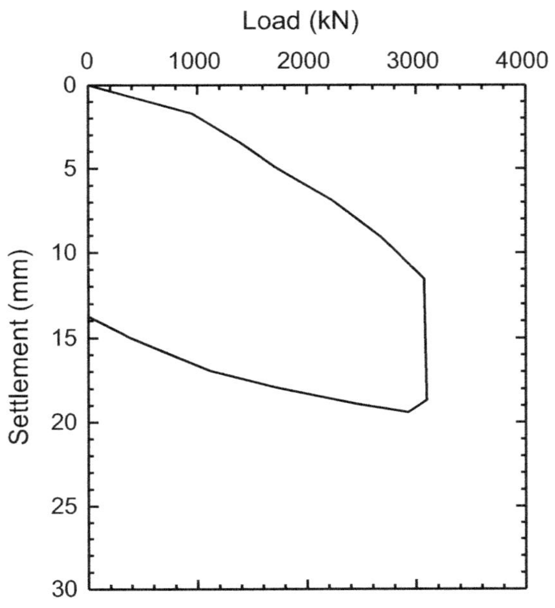

Detail	Description
Pile type/material	Square concrete pile
Length, L (m)	16.8
Outer width, B (mm)	610
Installation method	Driven
Set up time, days	12
Loading mode	Compression
$Q_{max\text{-}measured}$ (kN)	3095
Q_m (kN)	3095
Q_s (kN)	2406
Q_b (kN)	689
API Q_c (kN)	2820
Q_c/Q_m	0.91
UWA Q_c (kN)	2632
Q_c/Q_m	0.85
ICP Q_c (kN)	1913
Q_c/Q_m	0.62
Fugro Q_c (kN)	4007
Q_c/Q_m	1.29
NGI Q_c (kN)	3405
Q_c/Q_m	1.10

Site ID No. 10: Rotterdam Harbor, The Netherlands.

Ref.: Gijt et al. (1995): Comparison of statnamic load test and static load tests at the Rotterdam Harbour. First International Statnamic Seminar, Vancouver, 1995, 11p.

Cone penetrometer data

Detail	Description
Site name and location	Rotterdam Harbour
Soil type (s)	Silty sand and fine sand
Water table depth (m)	0
Pile type (s)	Square concrete pile
Type of cone penetrometer testing	CPT
Number of pile load tests	3
Comments	Altogether six piles. Three had enlarged toes and are therefore not accepted. Short CPT max-outs between 10 and 18m and no reliable q_c data below 26 m depth due to CPT capacity being exceeded. A value of 30 MPa is assumed below this level for analysis.

Pile ID: RH P6

Load–displacement data

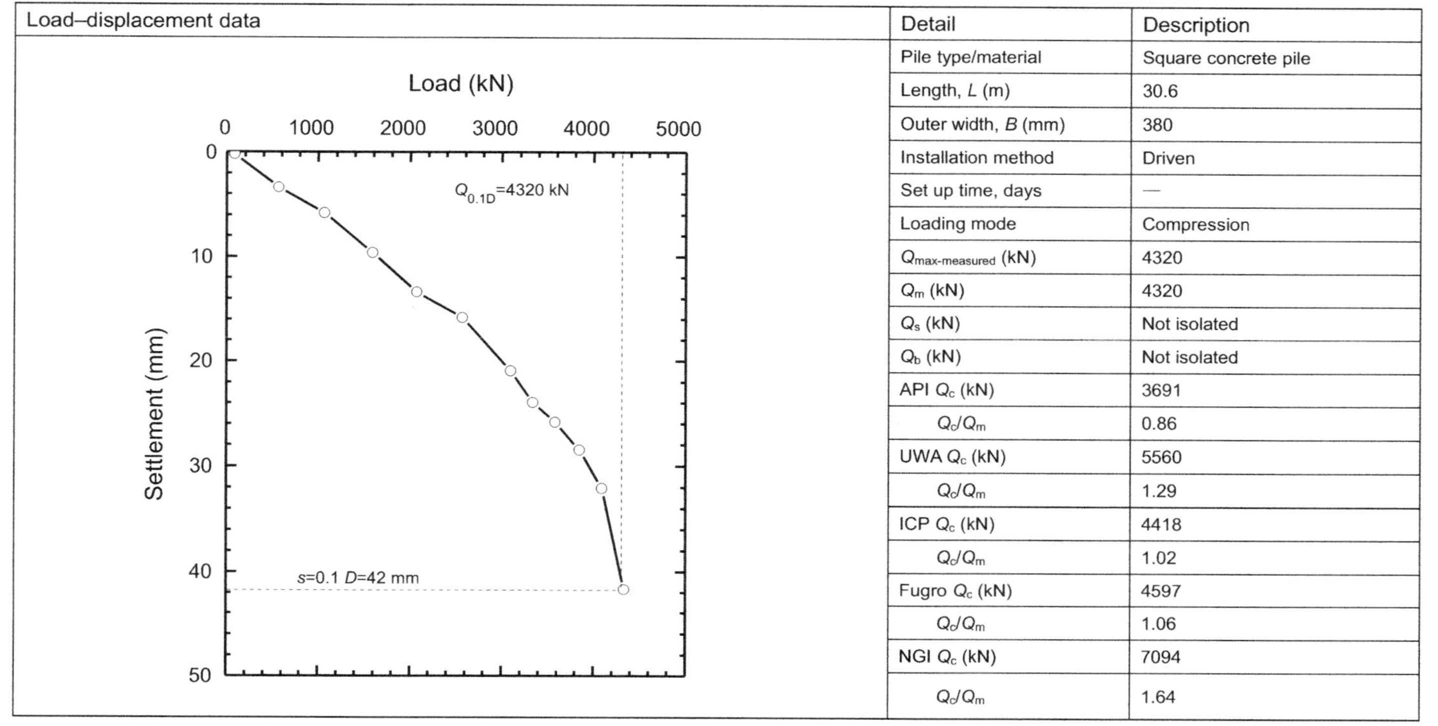

Detail	Description
Pile type/material	Square concrete pile
Length, L (m)	30.6
Outer width, B (mm)	380
Installation method	Driven
Set up time, days	—
Loading mode	Compression
$Q_{max\text{-}measured}$ (kN)	4320
Q_m (kN)	4320
Q_s (kN)	Not isolated
Q_b (kN)	Not isolated
API Q_c (kN)	3691
Q_c/Q_m	0.86
UWA Q_c (kN)	5560
Q_c/Q_m	1.29
ICP Q_c (kN)	4418
Q_c/Q_m	1.02
Fugro Q_c (kN)	4597
Q_c/Q_m	1.06
NGI Q_c (kN)	7094
Q_c/Q_m	1.64

Pile ID: RH P8

Load–displacement data

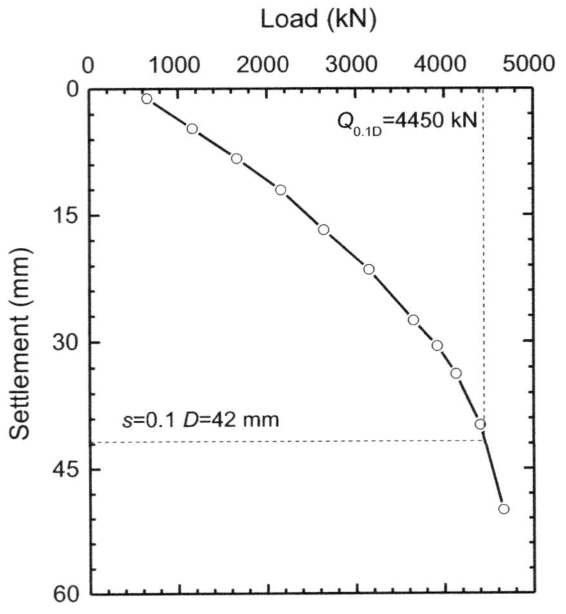

Detail	Description
Pile type/material	Square concrete pile
Length, L (m)	30.3
Outer width, B (mm)	380
Installation method	Driven
Set up time, days	—
Loading mode	Compression
$Q_{max\text{-}measured}$ (kN)	4650
Q_m (kN)	4450
Q_s (kN)	Not isolated
Q_b (kN)	Not isolated
API Q_c (kN)	3619
Q_c/Q_m	0.81
UWA Q_c (kN)	5491
Q_c/Q_m	1.23
ICP Q_c (kN)	4340
Q_c/Q_m	0.98
Fugro Q_c (kN)	4512
Q_c/Q_m	1.01
NGI Q_c (kN)	8394
Q_c/Q_m	1.89

Pile ID: RH P10

Load–displacement data

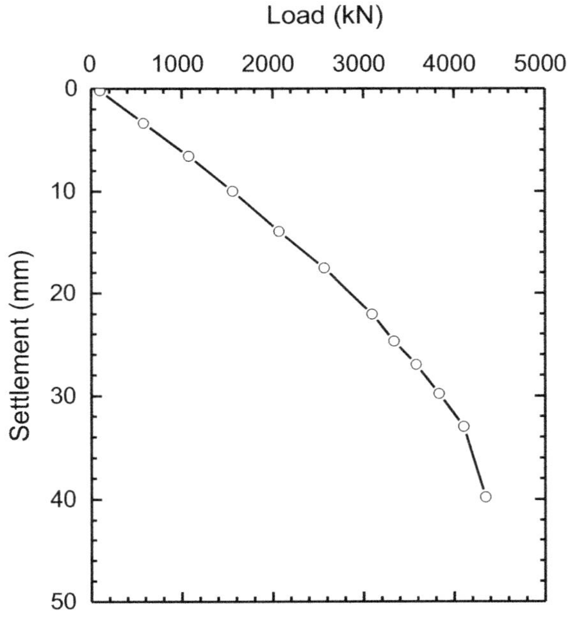

Detail	Description
Pile type/material	Square concrete pile
Length, L (m)	30.7
Outer width, B (mm)	380
Installation method	Driven
Set up time, days	—
Loading mode	Compression
$Q_{max\text{-}measured}$ (kN)	4330
Q_m (kN)	4330
Q_s (kN)	Not isolated
Q_b (kN)	Not isolated
API Q_c (kN)	3668
Q_c/Q_m	0.85
UWA Q_c (kN)	5579
Q_c/Q_m	1.29
ICP Q_c (kN)	4437
Q_c/Q_m	1.02
Fugro Q_c (kN)	4611
Q_c/Q_m	1.06
NGI Q_c (kN)	7126
Q_c/Q_m	1.65

Site ID No. 11: Waddinxveen Site, The Netherlands

Ref.: Holscher et al. (2009): Field test rapid load testing Waddinxveen. Deltares Factual report, The Netherlands, 95p.

Cone penetrometer data

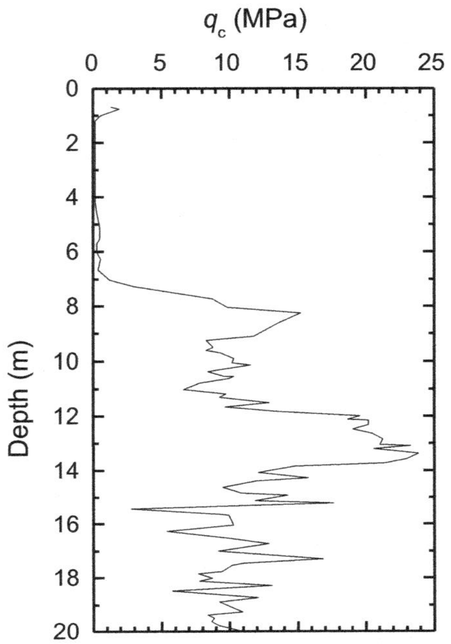

Detail	Description
Site name and location	Waddinxveen Site
Soil type (s)	Clay silt in the upper layer with fine sand followed
Water table depth (m)	10
Pile type (s)	Square concrete pile
Type of cone penetrometer testing	CPT
Number of pile load tests	1
Comments	Two piles tested, but one was tested statically and the other was subjected to Statnamic loading and has not been accepted. Interface friction angle and soil unit weight apply default value.

Pile ID: WDD P2

Load–displacement data

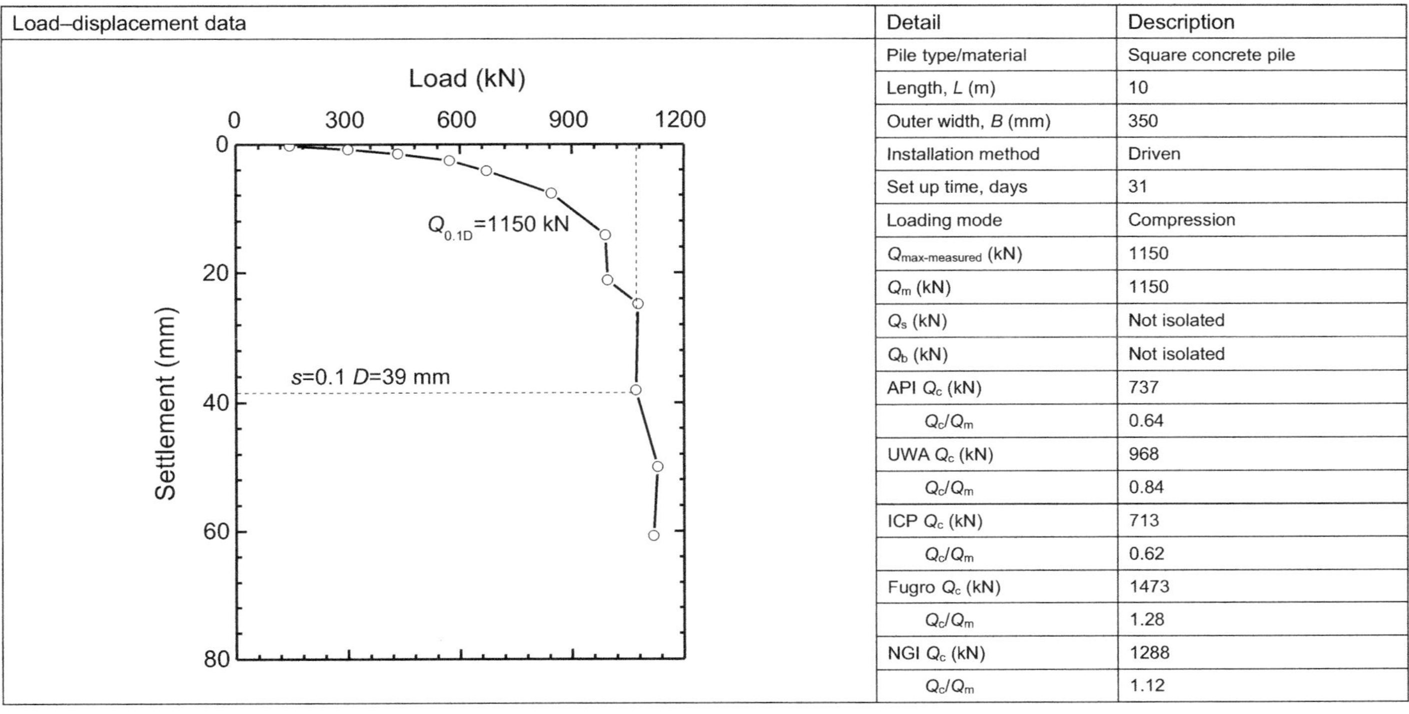

Detail	Description
Pile type/material	Square concrete pile
Length, L (m)	10
Outer width, B (mm)	350
Installation method	Driven
Set up time, days	31
Loading mode	Compression
$Q_{max\text{-}measured}$ (kN)	1150
Q_m (kN)	1150
Q_s (kN)	Not isolated
Q_b (kN)	Not isolated
API Q_c (kN)	737
Q_c/Q_m	0.64
UWA Q_c (kN)	968
Q_c/Q_m	0.84
ICP Q_c (kN)	713
Q_c/Q_m	0.62
Fugro Q_c (kN)	1473
Q_c/Q_m	1.28
NGI Q_c (kN)	1288
Q_c/Q_m	1.12

Site ID No. 12: Mobile Bay, AL, USA.

Ref.: Mayne (2013): FHWA Deep Foundation Load Test Database (DFLTD). Private communication.

Cone penetrometer data

q_c (MPa)
0 10 20 30 40
Depth (m)
0 5 10 15 20 25 30 35 40

f_s (kPa)
0 100 200 300 400 500
Depth (m)
0 5 10 15 20 25 30 35 40

Detail	Description
Site name and location	105GRL Piles-Mobile Bay, Mobile, AL Site
Soil type (s)	Silty sand and fine sand
Water table depth (m)	0
Pile type (s)	Open-ended steel piles
Type of cone penetrometer testing	CPT for AL1
Number of pile load tests	1
Comments	CPT profile is not very detailed, and given in tabular form only. Interface friction angle and soil unit weight apply default value.

Pile ID: AL 1

Load–displacement data

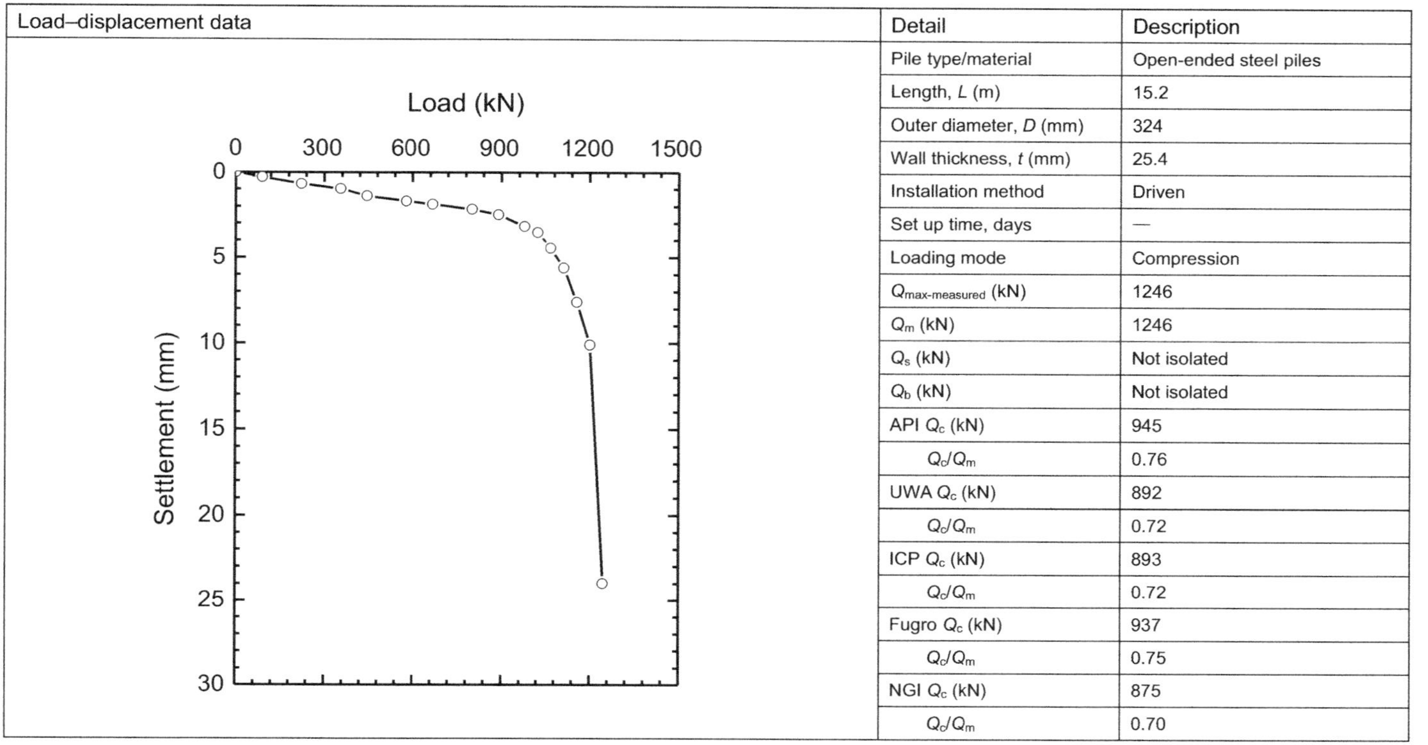

Detail	Description
Pile type/material	Open-ended steel piles
Length, L (m)	15.2
Outer diameter, D (mm)	324
Wall thickness, t (mm)	25.4
Installation method	Driven
Set up time, days	—
Loading mode	Compression
$Q_{max\text{-}measured}$ (kN)	1246
Q_m (kN)	1246
Q_s (kN)	Not isolated
Q_b (kN)	Not isolated
API Q_c (kN)	945
Q_c/Q_m	0.76
UWA Q_c (kN)	892
Q_c/Q_m	0.72
ICP Q_c (kN)	893
Q_c/Q_m	0.72
Fugro Q_c (kN)	937
Q_c/Q_m	0.75
NGI Q_c (kN)	875
Q_c/Q_m	0.70

Site ID No. 13: Mobile Bay, AL, USA.

Ref.: Mayne (2013): FHWA Deep Foundation Load Test Database (DFLTD). Private communication.

Cone penetrometer data

q_c (MPa) — Depth (m)

f_s (kPa) — Depth (m)

Detail	Description
Site name and location	105GRL Piles-Mobile Bay, Mobile, AL Site
Soil type (s)	Silty sand and fine sand
Water table depth (m)	0
Pile type (s)	Open-ended steel piles
Type of cone penetrometer testing	CPT for AL2
Number of pile load tests	1
Comments	CPT profile is not very detailed, and given in tabular form only. Interface friction angle and soil unit weight apply default value.

Pile ID: AL 2

Load–displacement data

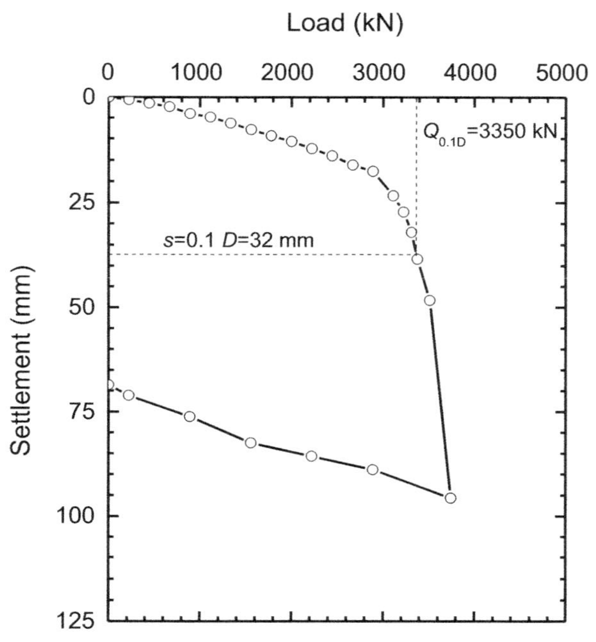

Detail	Description
Pile type/material	Open-ended steel piles
Length, L (m)	42.7
Outer diameter, D (mm)	324
Wall thickness, t (mm)	25.4
Installation method	Driven
Set up time, days	—
Loading mode	Compression
$Q_{max\text{-}measured}$ (kN)	3750
Q_m (kN)	3350
Q_s (kN)	Not isolated
Q_b (kN)	Not isolated
API Q_c (kN)	2618
Q_c/Q_m	0.78
UWA Q_c (kN)	2636
Q_c/Q_m	0.79
ICP Q_c (kN)	3093
Q_c/Q_m	0.92
Fugro Q_c (kN)	1912
Q_c/Q_m	0.57
NGI Q_c (kN)	2820
Q_c/Q_m	0.84

Site ID No. 14: ABEF Foundation, Brazil.

Ref.: Mayne (2013): FHWA Deep Foundation Load Test Database (DFLTD). Private communication.

Cone penetrometer data

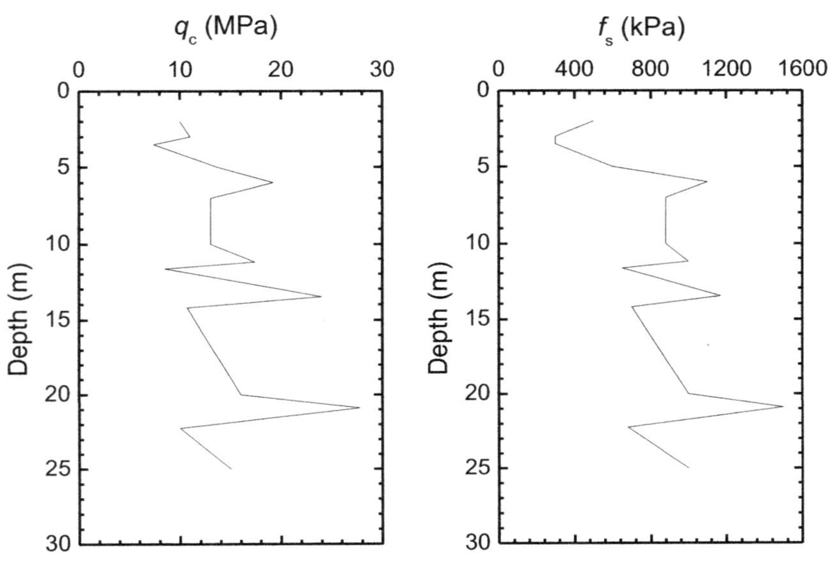

Detail	Description
Site name and location	84 ABEF Research on Foundation # 84 Site
Soil type (s)	Silty sand and fine sand
Water table depth (m)	42.6
Pile type (s)	Open-ended concrete piles
Type of cone penetrometer testing	CPT for pile NA1
Number of pile load tests	1
Comments	CPT profile is not very detailed, and given in tabular form only. Interface friction angle and soil unit weight apply default value.

Pile ID: NA 1

Load–displacement data

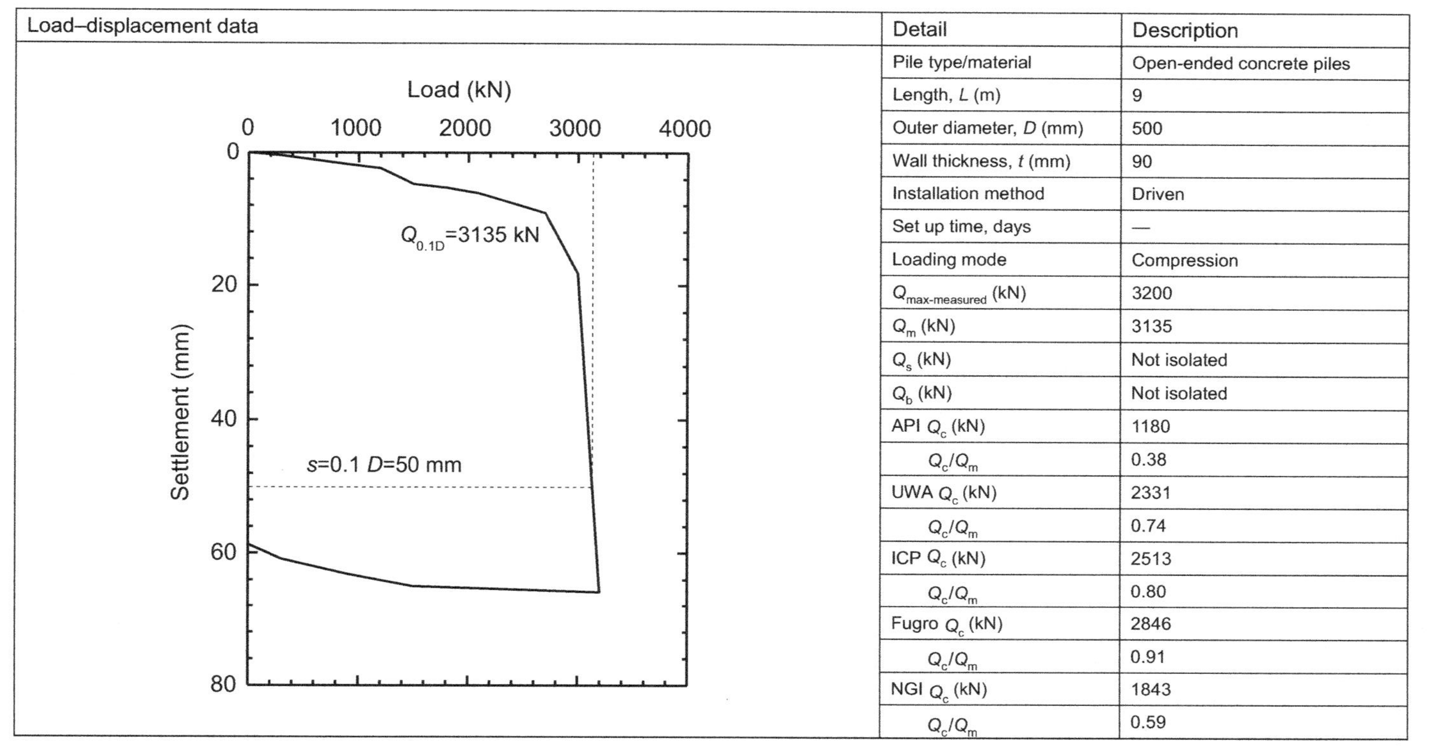

Detail	Description
Pile type/material	Open-ended concrete piles
Length, L (m)	9
Outer diameter, D (mm)	500
Wall thickness, t (mm)	90
Installation method	Driven
Set up time, days	—
Loading mode	Compression
$Q_{max\text{-}measured}$ (kN)	3200
Q_m (kN)	3135
Q_s (kN)	Not isolated
Q_b (kN)	Not isolated
API Q_c (kN)	1180
Q_c/Q_m	0.38
UWA Q_c (kN)	2331
Q_c/Q_m	0.74
ICP Q_c (kN)	2513
Q_c/Q_m	0.80
Fugro Q_c (kN)	2846
Q_c/Q_m	0.91
NGI Q_c (kN)	1843
Q_c/Q_m	0.59

Site ID No. 15: ABEF Foundation, Brazil.

Ref.: Mayne (2013): FHWA Deep Foundation Load Test Database (DFLTD). Private communication.

Cone penetrometer data

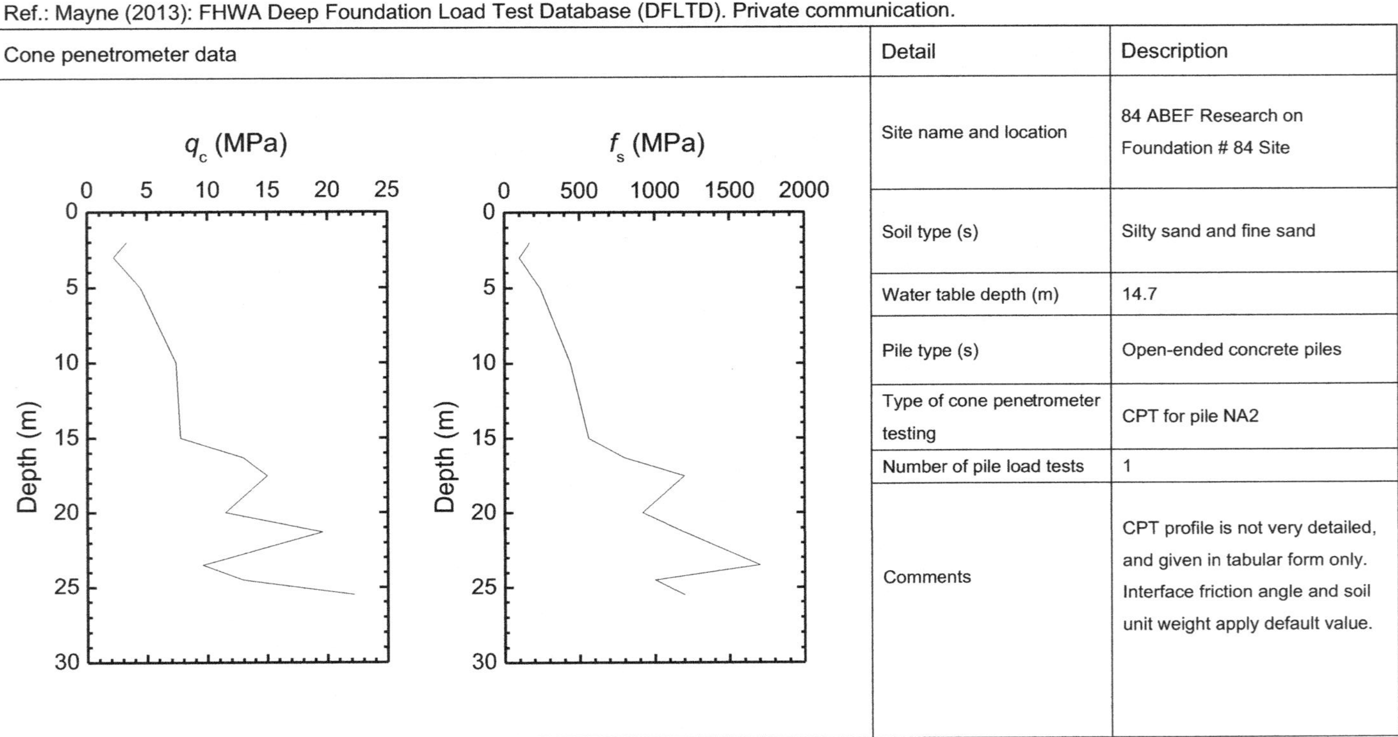

Detail	Description
Site name and location	84 ABEF Research on Foundation # 84 Site
Soil type (s)	Silty sand and fine sand
Water table depth (m)	14.7
Pile type (s)	Open-ended concrete piles
Type of cone penetrometer testing	CPT for pile NA2
Number of pile load tests	1
Comments	CPT profile is not very detailed, and given in tabular form only. Interface friction angle and soil unit weight apply default value.

Pile ID: NA 2

Load–displacement data

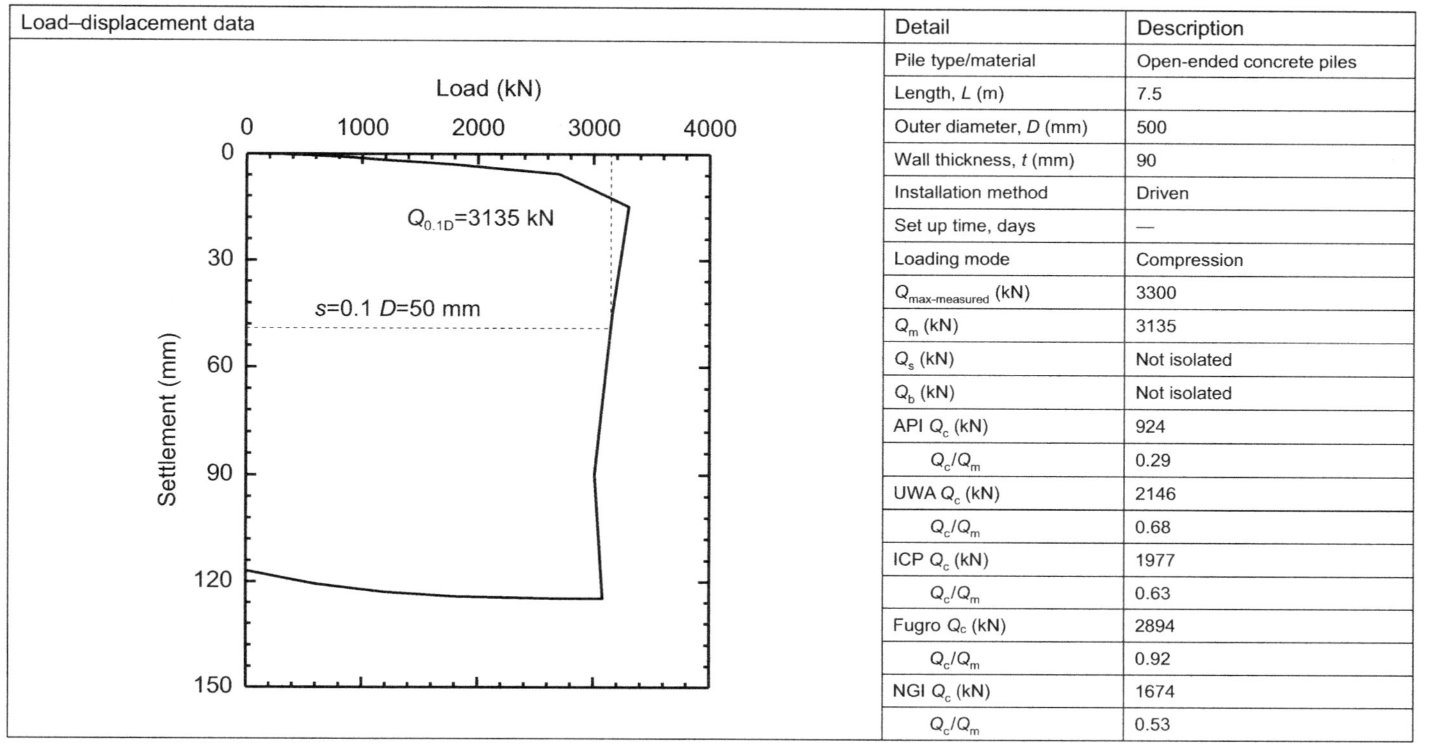

Detail	Description
Pile type/material	Open-ended concrete piles
Length, L (m)	7.5
Outer diameter, D (mm)	500
Wall thickness, t (mm)	90
Installation method	Driven
Set up time, days	—
Loading mode	Compression
$Q_{max\text{-}measured}$ (kN)	3300
Q_m (kN)	3135
Q_s (kN)	Not isolated
Q_b (kN)	Not isolated
API Q_c (kN)	924
Q_c/Q_m	0.29
UWA Q_c (kN)	2146
Q_c/Q_m	0.68
ICP Q_c (kN)	1977
Q_c/Q_m	0.63
Fugro Q_c (kN)	2894
Q_c/Q_m	0.92
NGI Q_c (kN)	1674
Q_c/Q_m	0.53

Site ID No. 16: Apalachicola River, USA.

Ref.: Mayne (2013): FHWA Deep Foundation Load Test Database (DFLTD). Private communication.

Cone penetrometer data

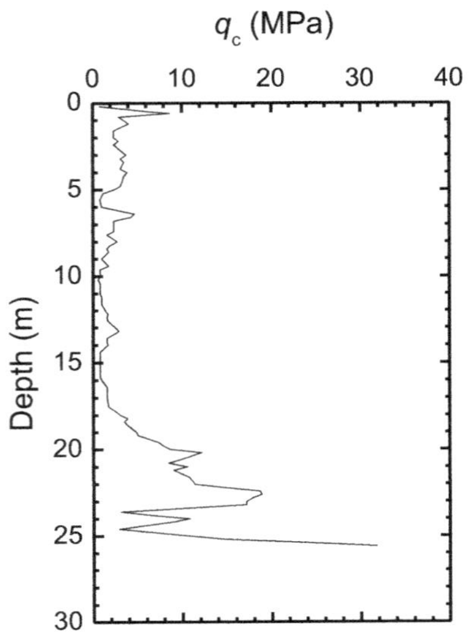

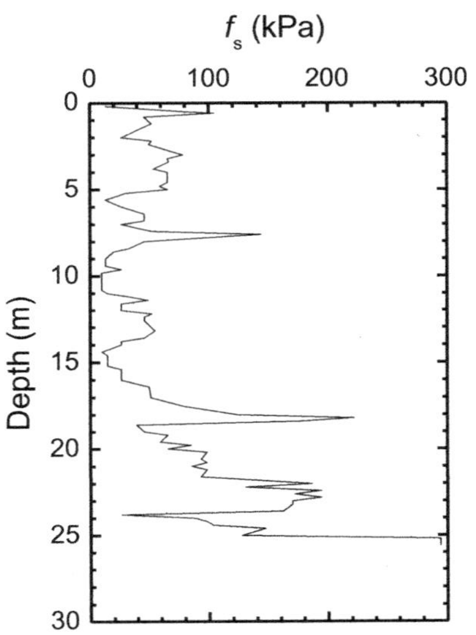

Detail	Description
Site name and location	1 Grl Piles-Apalachicola River Br SITE
Soil type (s)	Silty sand and fine sand
Water table depth (m)	0
Pile type (s)	Square concrete piles
Type of cone penetrometer testing	CPT
Number of pile load tests	1
Comments	These tests were conducted at a silica sand location and these are not the FDOT cemented sand site tests. Interface friction angle and soil unit weight apply default value.

Pile ID: BR 1

Load–displacement data

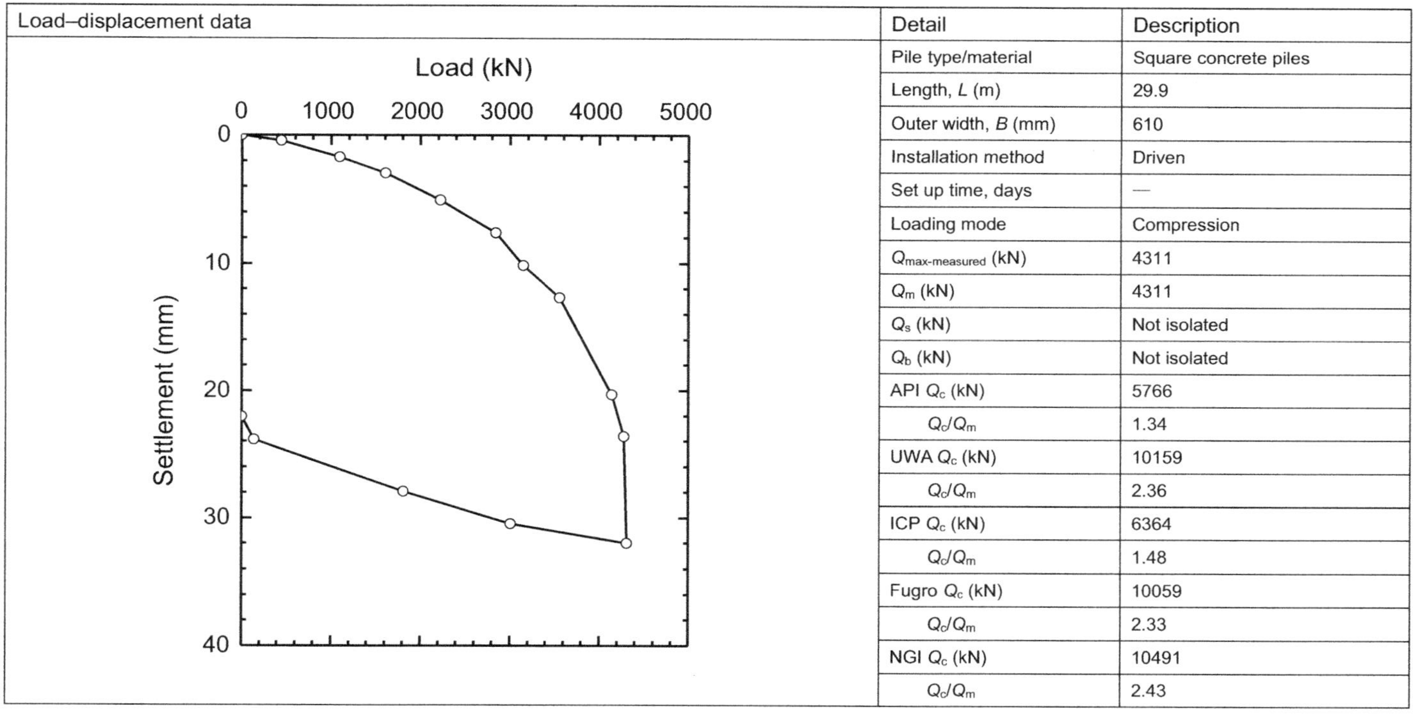

Detail	Description
Pile type/material	Square concrete piles
Length, L (m)	29.9
Outer width, B (mm)	610
Installation method	Driven
Set up time, days	—
Loading mode	Compression
$Q_{max\text{-}measured}$ (kN)	4311
Q_m (kN)	4311
Q_s (kN)	Not isolated
Q_b (kN)	Not isolated
API Q_c (kN)	5766
Q_c/Q_m	1.34
UWA Q_c (kN)	10159
Q_c/Q_m	2.36
ICP Q_c (kN)	6364
Q_c/Q_m	1.48
Fugro Q_c (kN)	10059
Q_c/Q_m	2.33
NGI Q_c (kN)	10491
Q_c/Q_m	2.43

Site ID No. 17: Los Angeles, CA Site, USA.

Ref.: Mayne (2013): FHWA Deep Foundation Load Test Database (DFLTD). Private communication.

Cone penetrometer data

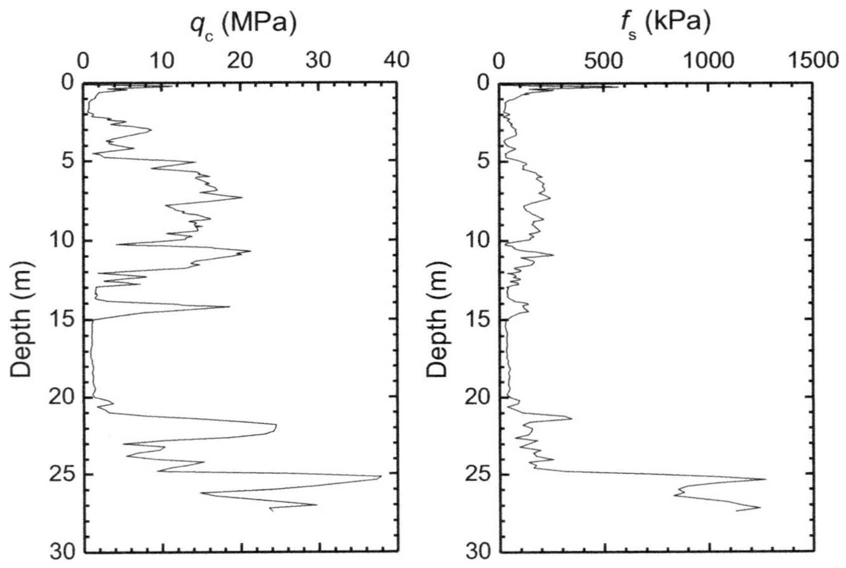

Detail	Description
Site name and location	129 GRL Piles-Port of Los Angeles, CA Site
Soil type (s)	Silty sand and fine sand
Water table depth (m)	0
Pile type (s)	Square concrete piles
Type of cone penetrometer testing	CPT
Number of pile load tests	1
Comments	Interface friction angle and soil unit weight apply default value.

Pile ID: CA 1

Load–displacement data

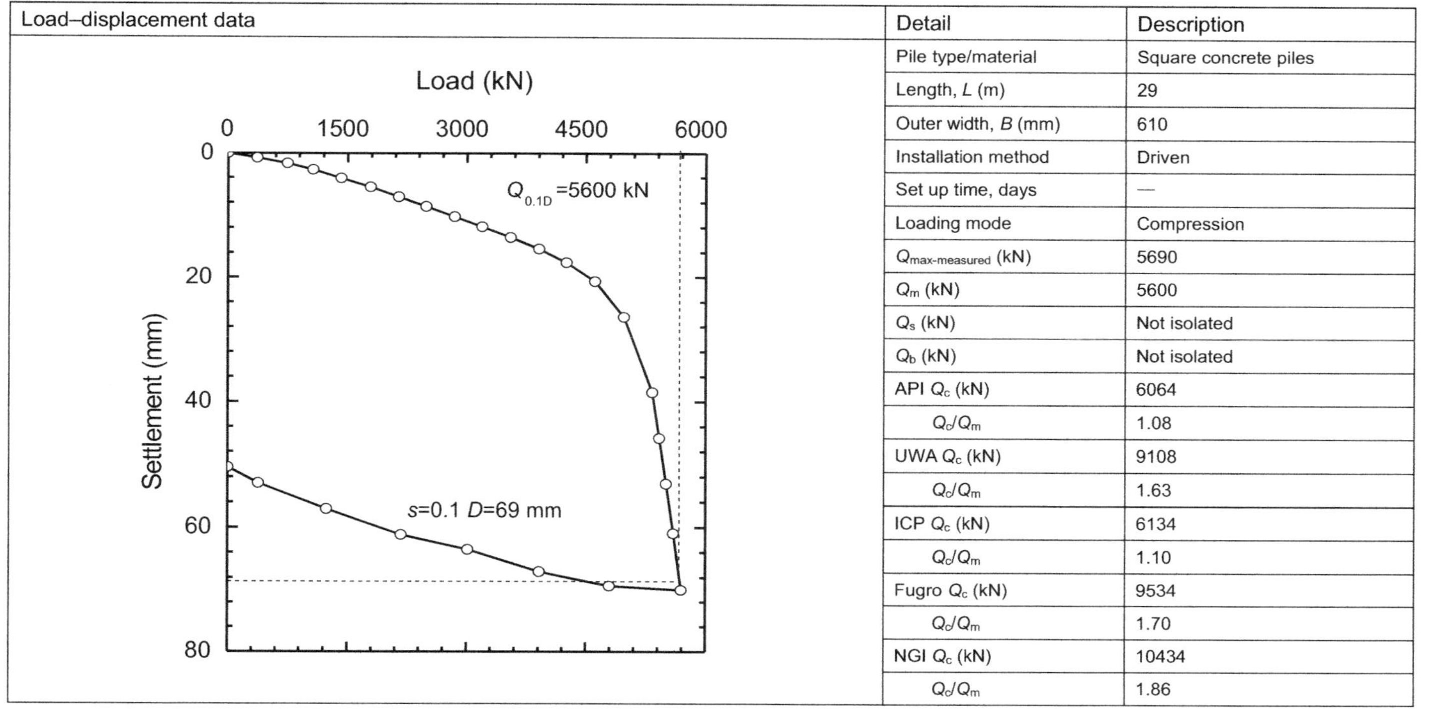

Detail	Description
Pile type/material	Square concrete piles
Length, L (m)	29
Outer width, B (mm)	610
Installation method	Driven
Set up time, days	—
Loading mode	Compression
$Q_{max\text{-}measured}$ (kN)	5690
Q_m (kN)	5600
Q_s (kN)	Not isolated
Q_b (kN)	Not isolated
API Q_c (kN)	6064
Q_c/Q_m	1.08
UWA Q_c (kN)	9108
Q_c/Q_m	1.63
ICP Q_c (kN)	6134
Q_c/Q_m	1.10
Fugro Q_c (kN)	9534
Q_c/Q_m	1.70
NGI Q_c (kN)	10434
Q_c/Q_m	1.86

Site ID No. 18: MS Smith, USA.

Ref.: Mayne (2013): FHWA Deep Foundation Load Test Database (DFLTD). Private communication.

Cone penetrometer data

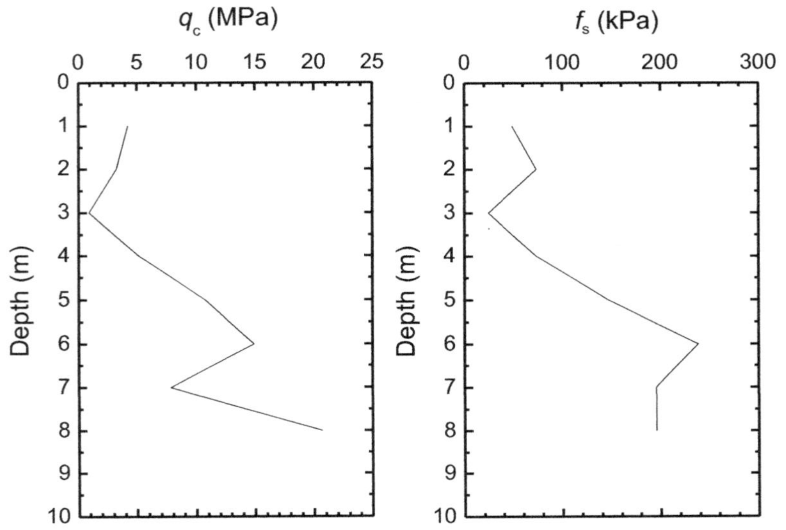

Detail	Description
Site name and location	Bridge Site 1045 MS Smith
Soil type (s)	Silty sand and fine sand
Water table depth (m)	0
Pile type (s)	Square concrete piles
Type of cone penetrometer testing	CPT
Number of pile load tests	1
Comments	CPT profile is not very detailed, and given in tabular form only. Interface friction angle and soil unit weight apply default value.

Pile ID: BR 1045

Load–displacement data

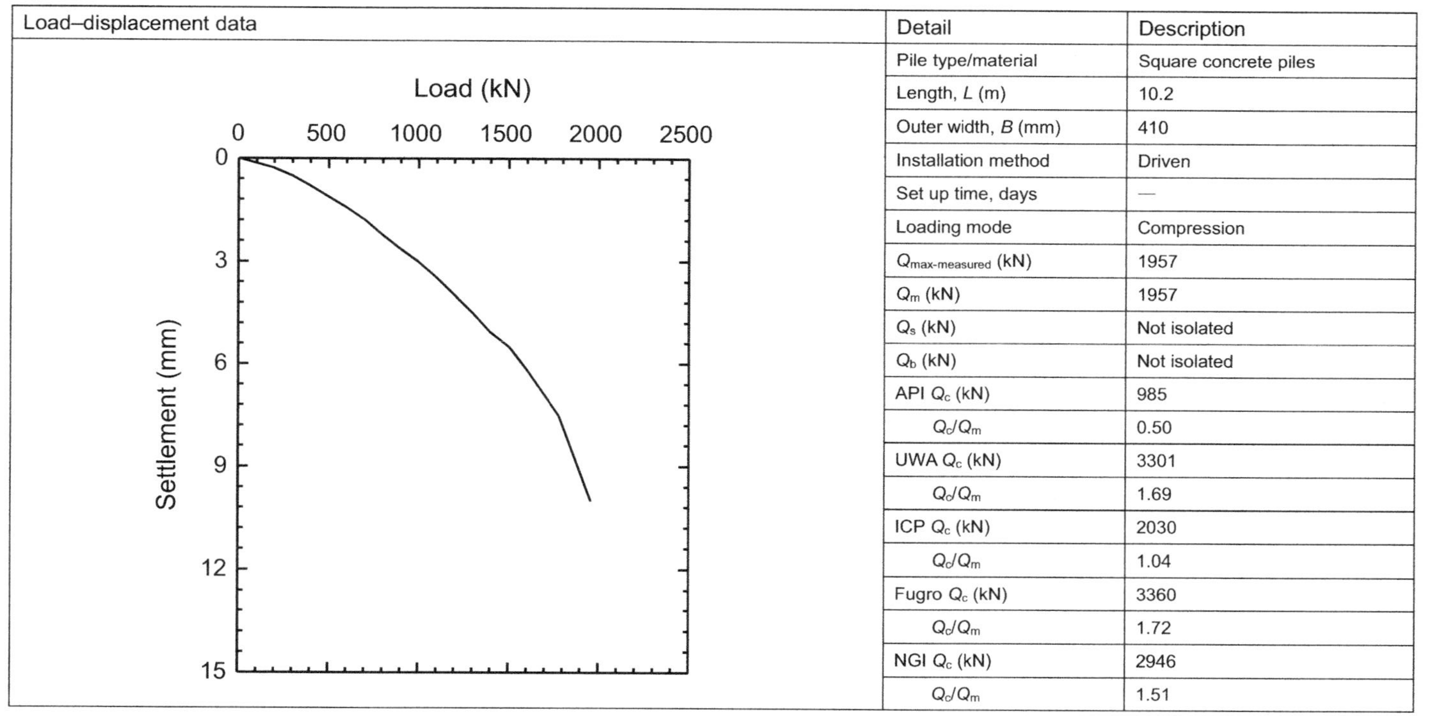

Detail	Description
Pile type/material	Square concrete piles
Length, L (m)	10.2
Outer width, B (mm)	410
Installation method	Driven
Set up time, days	—
Loading mode	Compression
$Q_{max\text{-}measured}$ (kN)	1957
Q_m (kN)	1957
Q_s (kN)	Not isolated
Q_b (kN)	Not isolated
API Q_c (kN)	985
Q_c/Q_m	0.50
UWA Q_c (kN)	3301
Q_c/Q_m	1.69
ICP Q_c (kN)	2030
Q_c/Q_m	1.04
Fugro Q_c (kN)	3360
Q_c/Q_m	1.72
NGI Q_c (kN)	2946
Q_c/Q_m	1.51

Site ID No. 19: MS Desota, USA.

Ref.: Mayne (2013): FHWA Deep Foundation Load Test Database (DFLTD). Private communication.

Cone penetrometer data	Detail	Description
	Site name and location	Bridge Site 2108 MS Desota
	Soil type (s)	Silty sand and fine sand
	Water table depth (m)	0
	Pile type (s)	Square concrete piles
	Type of cone penetrometer testing	CPT
	Number of pile load tests	1
	Comments	CPT profile is not very detailed, and given in tabular form only. Interface friction angle and soil unit weight apply default value.

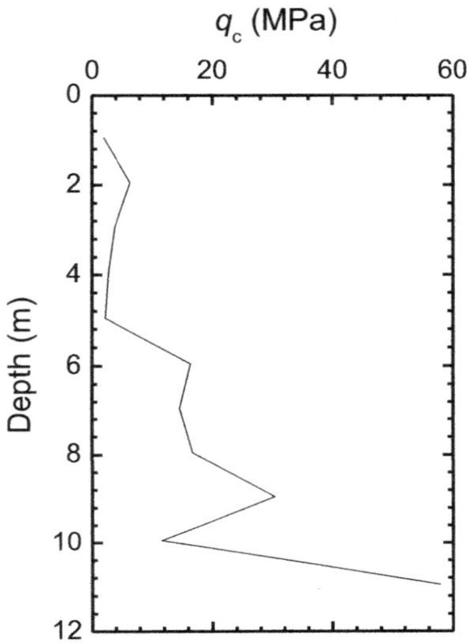

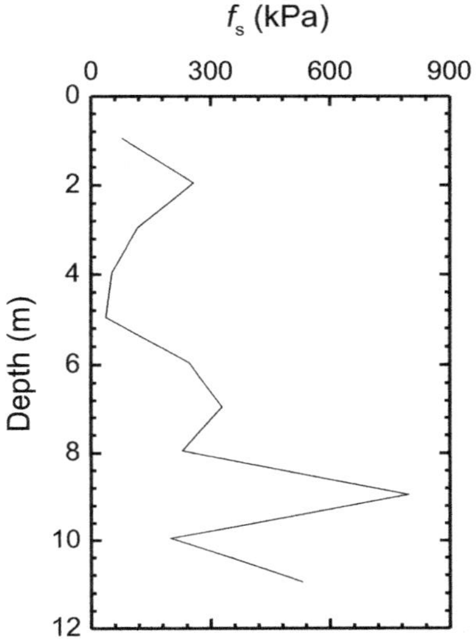

Pile ID: BR 2108

Load–displacement data

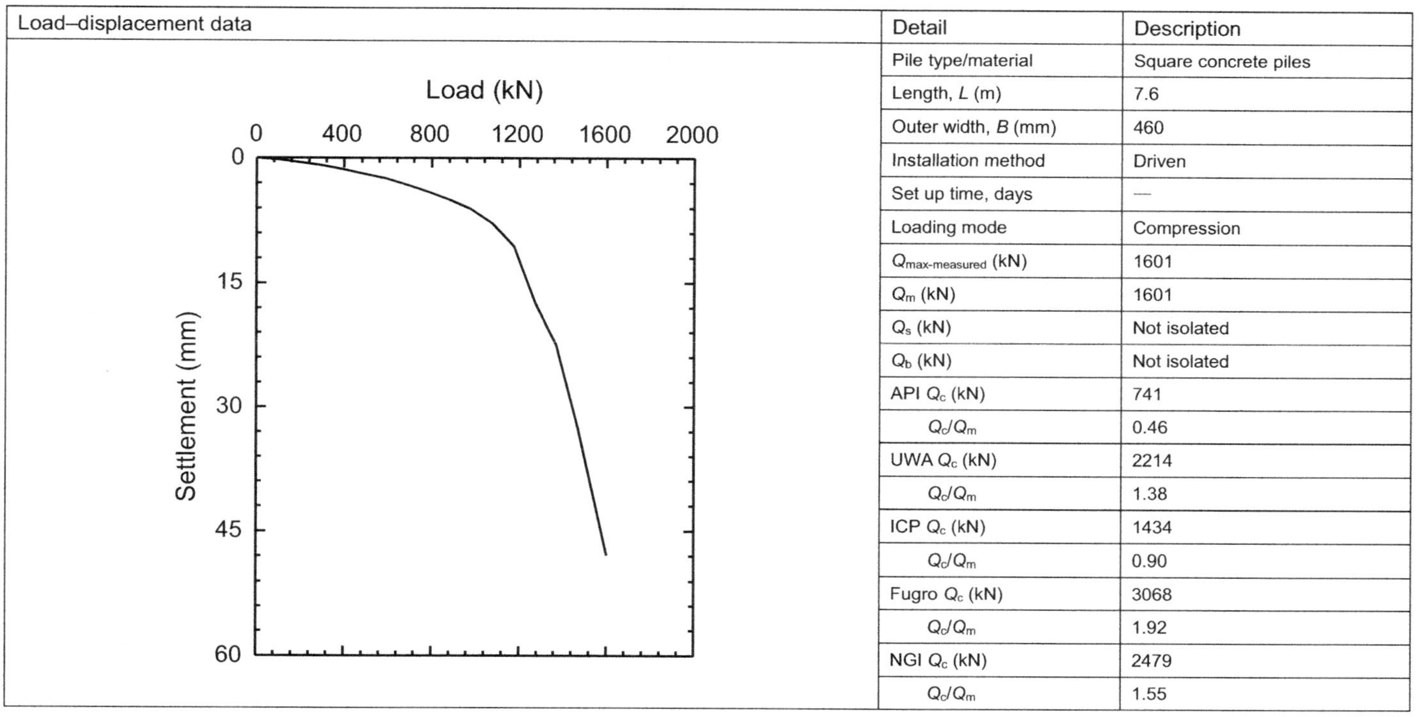

Detail	Description
Pile type/material	Square concrete piles
Length, L (m)	7.6
Outer width, B (mm)	460
Installation method	Driven
Set up time, days	—
Loading mode	Compression
$Q_{max\text{-}measured}$ (kN)	1601
Q_m (kN)	1601
Q_s (kN)	Not isolated
Q_b (kN)	Not isolated
API Q_c (kN)	741
Q_c/Q_m	0.46
UWA Q_c (kN)	2214
Q_c/Q_m	1.38
ICP Q_c (kN)	1434
Q_c/Q_m	0.90
Fugro Q_c (kN)	3068
Q_c/Q_m	1.92
NGI Q_c (kN)	2479
Q_c/Q_m	1.55

Site ID No. 20: MS Harrison, USA.

Ref.: Mayne (2013): FHWA Deep Foundation Load Test Database (DFLTD). Private communication.

Cone penetrometer data

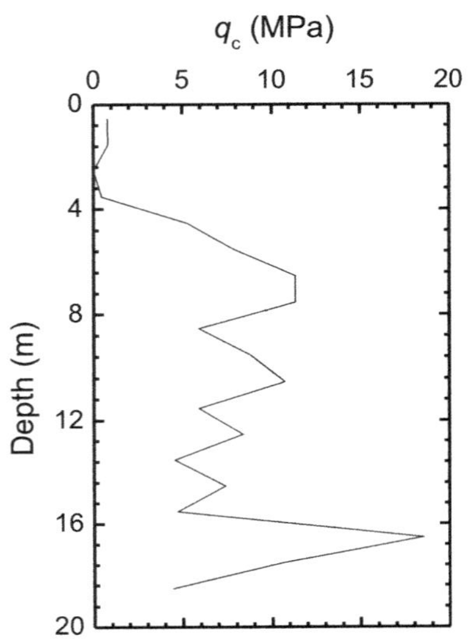

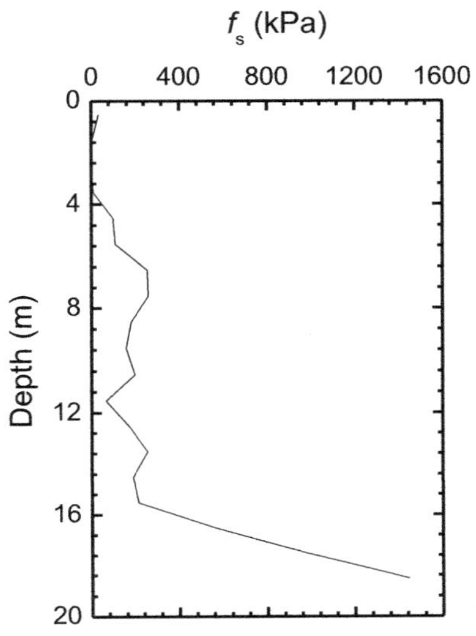

Detail	Description
Site name and location	Bridge Site 3028 MS Harrison
Soil type (s)	Silty sand and fine sand
Water table depth (m)	0
Pile type (s)	Square concrete piles
Type of cone penetrometer testing	CPT
Number of pile load tests	1
Comments	CPT profile is not very detailed, and given in tabular form only. Interface friction angle and soil unit weight apply default value.

Pile ID: BR 3028

Load–displacement data

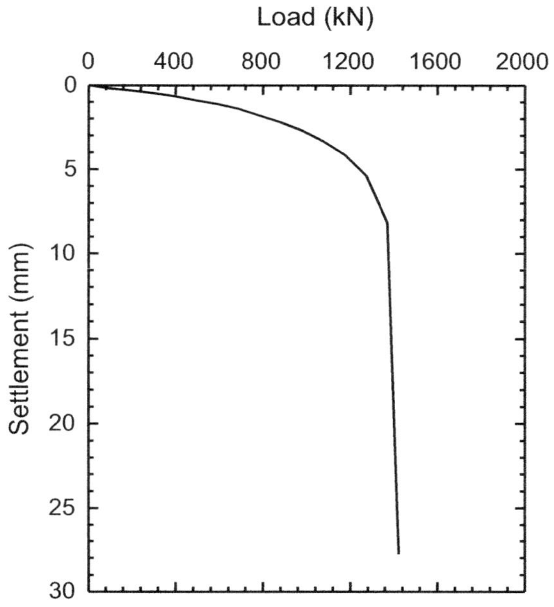

Detail	Description
Pile type/material	Square concrete piles
Length, L (m)	7.6
Outer width, B (mm)	460
Installation method	Driven
Set up time, days	—
Loading mode	Compression
$Q_{max\text{-}measured}$ (kN)	1425
Q_m (kN)	1425
Q_s (kN)	Not isolated
Q_b (kN)	Not isolated
API Q_c (kN)	2081
Q_c/Q_m	1.46
UWA Q_c (kN)	2177
Q_c/Q_m	1.53
ICP Q_c (kN)	1612
Q_c/Q_m	1.13
Fugro Q_c (kN)	2770
Q_c/Q_m	1.94
NGI Q_c (kN)	3318
Q_c/Q_m	2.33

Site ID No. 21: Washington MS, USA.

Ref.: Mayne (2013): FHWA Deep Foundation Load Test Database (DFLTD). Private communication.

Cone penetrometer data

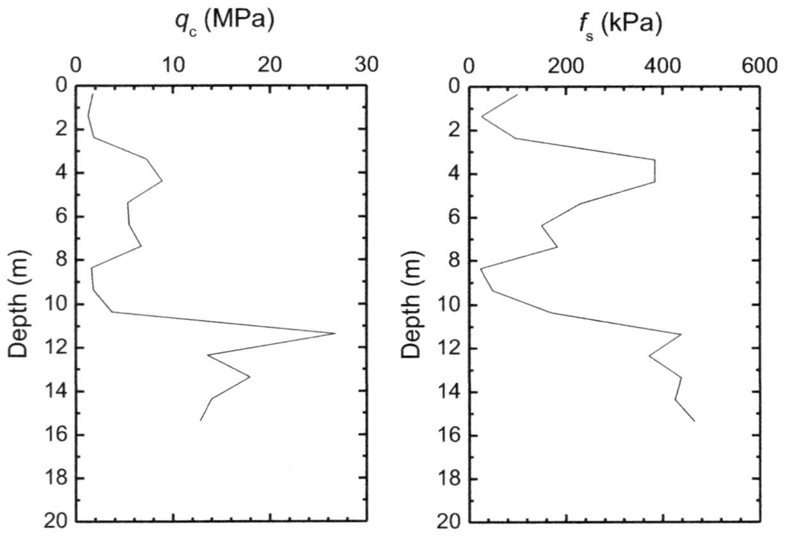

Detail	Description
Site name and location	Bridge Site 3118A Washington
Soil type (s)	Silty sand and fine sand
Water table depth (m)	0
Pile type (s)	Square concrete piles
Type of cone penetrometer testing	CPT
Number of pile load tests	1
Comments	CPT profile is not very detailed, and given in tabular form only. Interface friction angle and soil unit weight apply default value.

Pile ID: BR 3118A

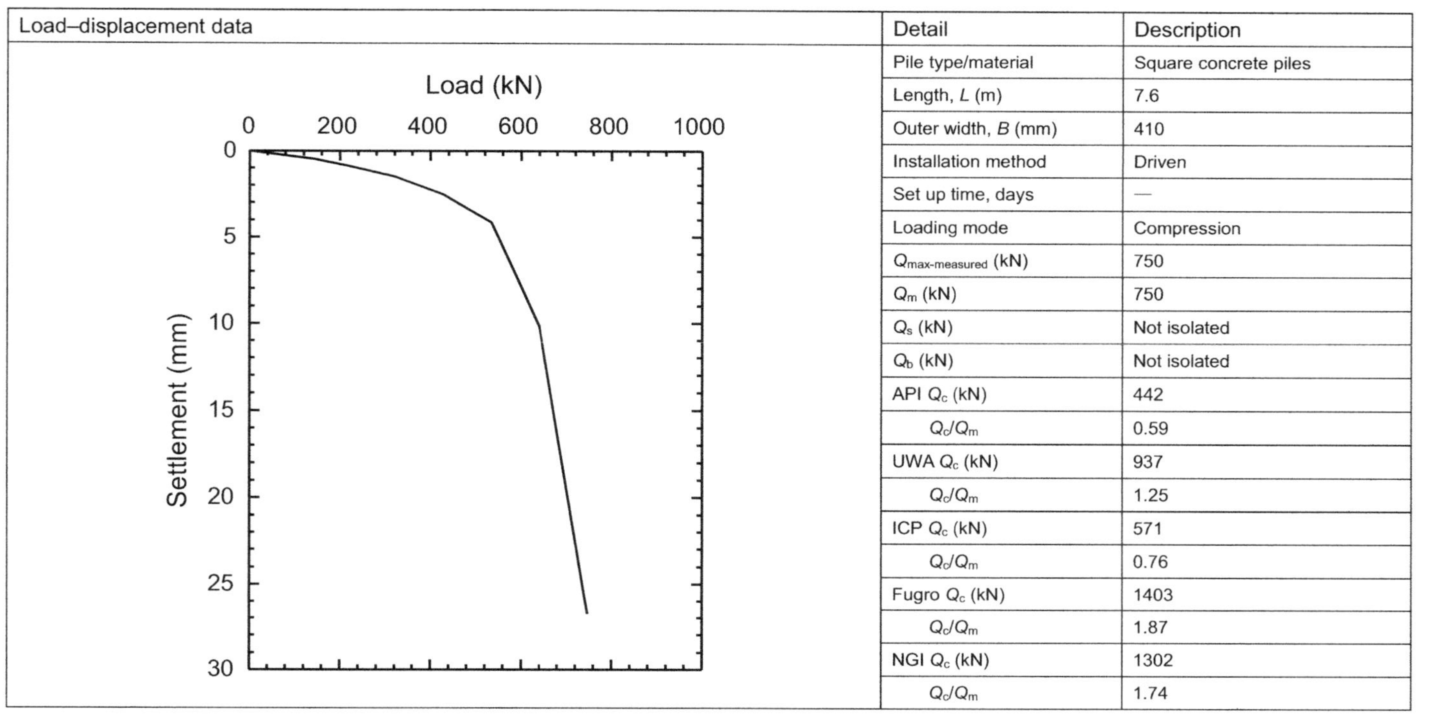

Detail	Description
Pile type/material	Square concrete piles
Length, L (m)	7.6
Outer width, B (mm)	410
Installation method	Driven
Set up time, days	—
Loading mode	Compression
$Q_{max\text{-}measured}$ (kN)	750
Q_m (kN)	750
Q_s (kN)	Not isolated
Q_b (kN)	Not isolated
API Q_c (kN)	442
Q_c/Q_m	0.59
UWA Q_c (kN)	937
Q_c/Q_m	1.25
ICP Q_c (kN)	571
Q_c/Q_m	0.76
Fugro Q_c (kN)	1403
Q_c/Q_m	1.87
NGI Q_c (kN)	1302
Q_c/Q_m	1.74

Site ID No. 22: Washington MS, USA.

Ref.: Mayne (2013): FHWA Deep Foundation Load Test Database (DFLTD). Private communication.

Cone penetrometer data

Detail	Description
Site name and location	Bridge Site 3123B Washington
Soil type (s)	Silty sand and fine sand
Water table depth (m)	0
Pile type (s)	Square concrete piles
Type of cone penetrometer testing	CPT
Number of pile load tests	1
Comments	CPT profile is not very detailed, and given in tabular form only. Interface friction angle and soil unit weight apply default value.

Pile ID: BR 3123B

Load–displacement data

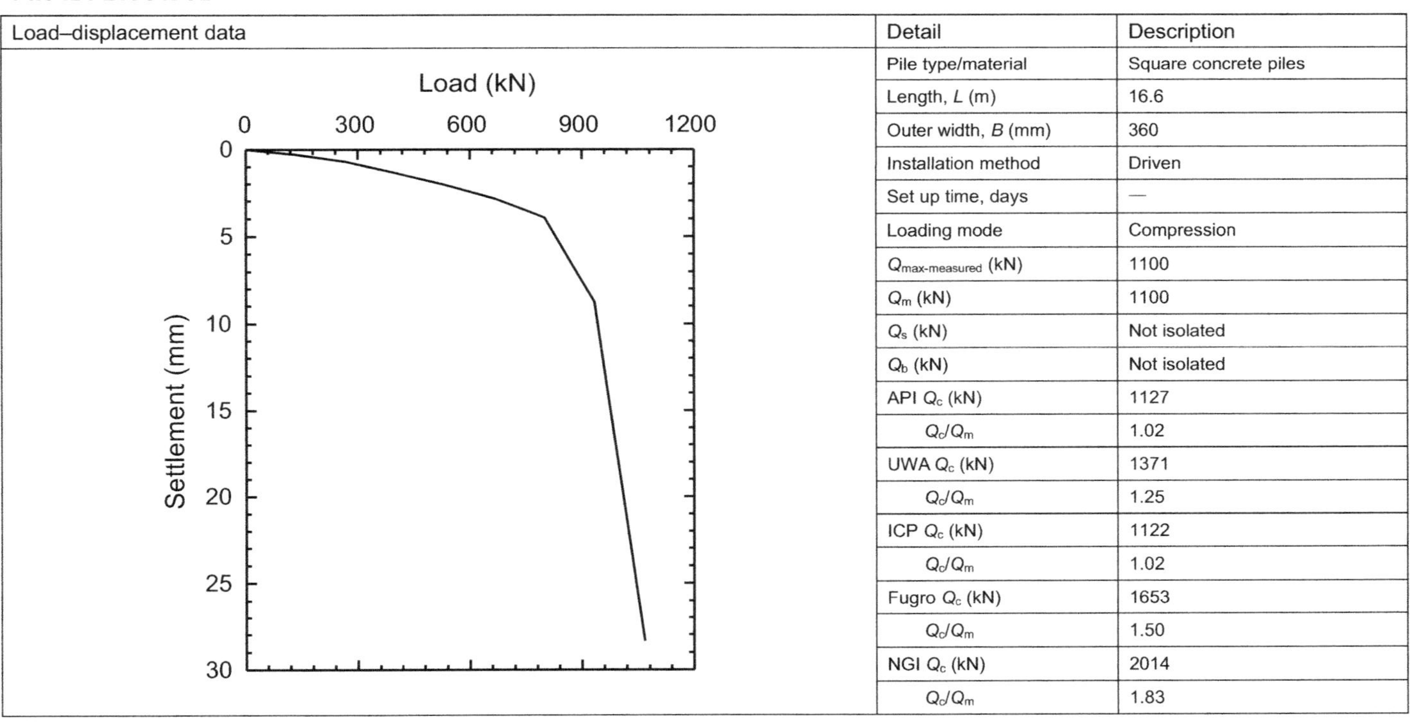

Detail	Description
Pile type/material	Square concrete piles
Length, L (m)	16.6
Outer width, B (mm)	360
Installation method	Driven
Set up time, days	—
Loading mode	Compression
$Q_{max\text{-}measured}$ (kN)	1100
Q_m (kN)	1100
Q_s (kN)	Not isolated
Q_b (kN)	Not isolated
API Q_c (kN)	1127
Q_c/Q_m	1.02
UWA Q_c (kN)	1371
Q_c/Q_m	1.25
ICP Q_c (kN)	1122
Q_c/Q_m	1.02
Fugro Q_c (kN)	1653
Q_c/Q_m	1.50
NGI Q_c (kN)	2014
Q_c/Q_m	1.83

Site ID No. 23: Washington MS, USA.

Ref.: Mayne (2013): FHWA Deep Foundation Load Test Database (DFLTD). Private communication.

Cone penetrometer data

q_c (MPa)

Depth (m)

f_s (kPa)

Depth (m)

Detail	Description
Site name and location	Bridge Site 3142A Washington
Soil type (s)	Silty sand and fine sand
Water table depth (m)	0
Pile type (s)	Square concrete piles
Type of cone penetrometer testing	CPT
Number of pile load tests	1
Comments	CPT profile is not very detailed, given in tabular form only. Interface friction angle and soil unit weight apply default value.

Pile ID: BR 3142A

Load–displacement data

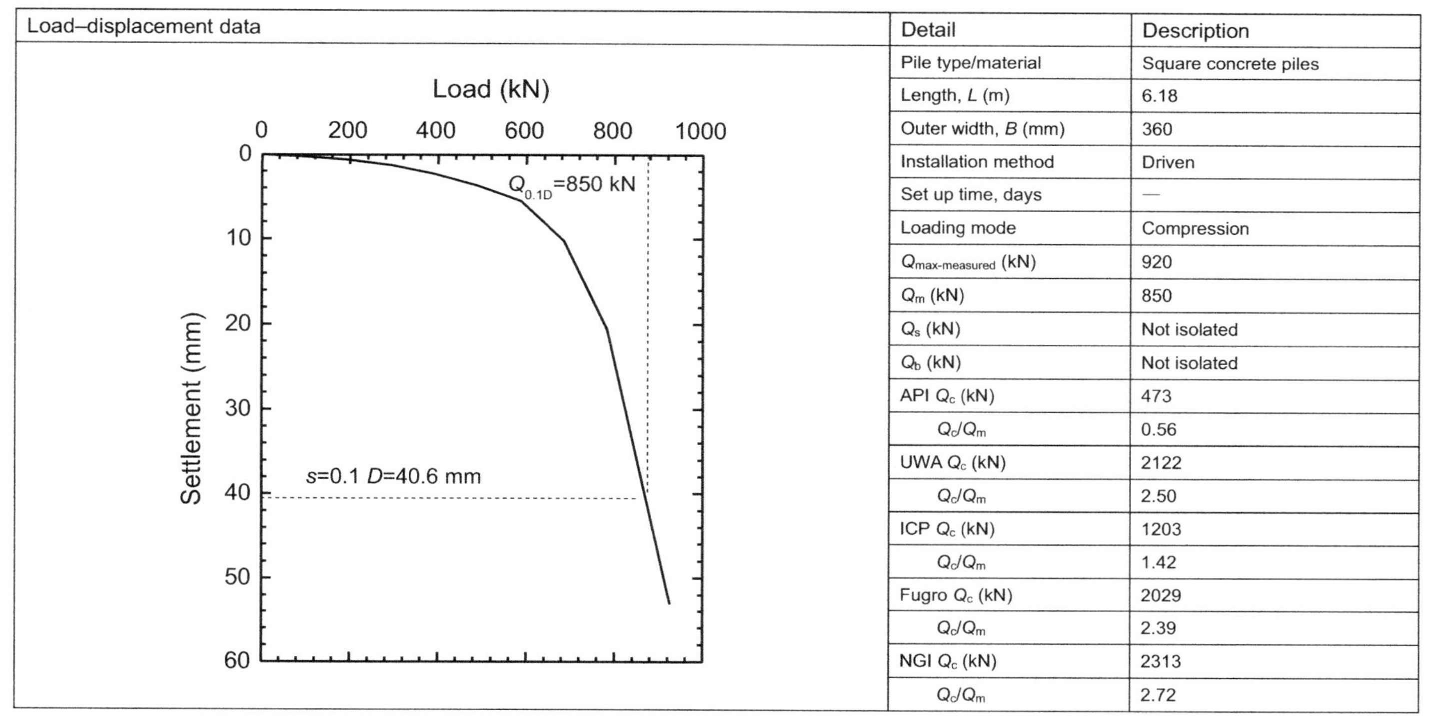

Detail	Description
Pile type/material	Square concrete piles
Length, L (m)	6.18
Outer width, B (mm)	360
Installation method	Driven
Set up time, days	—
Loading mode	Compression
$Q_{max\text{-}measured}$ (kN)	920
Q_m (kN)	850
Q_s (kN)	Not isolated
Q_b (kN)	Not isolated
API Q_c (kN)	473
Q_c/Q_m	0.56
UWA Q_c (kN)	2122
Q_c/Q_m	2.50
ICP Q_c (kN)	1203
Q_c/Q_m	1.42
Fugro Q_c (kN)	2029
Q_c/Q_m	2.39
NGI Q_c (kN)	2313
Q_c/Q_m	2.72

Site ID No. 24: Larvik, Norway.

Ref.: Karlsrud (2013): Summary and evaluation of pile test results. NGI Report: Time effects on pile capacity. Doi: 20061251-00-279-R.

Cone penetrometer data	Detail	Description
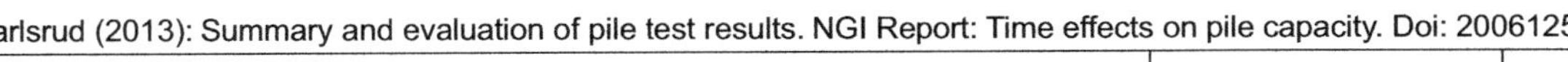	Site name and location	Larvik
	Soil type (s)	Silty sand and fine sand
	Water table depth (m)	2
	Pile type (s)	Open-ended steel pile
	Type of cone penetrometer testing	CPT
	Number of pile load tests	7
	Comments	A set of seven piles subjected to staged tension tests over an extended period. Some piles were allowed to age under constant tension loading. Only first time, previously unloaded piles included here. Interface friction angle applies default value. Soil unit weight applies the values described in the report.

Pile ID: Larvik L1-1

Load–displacement data

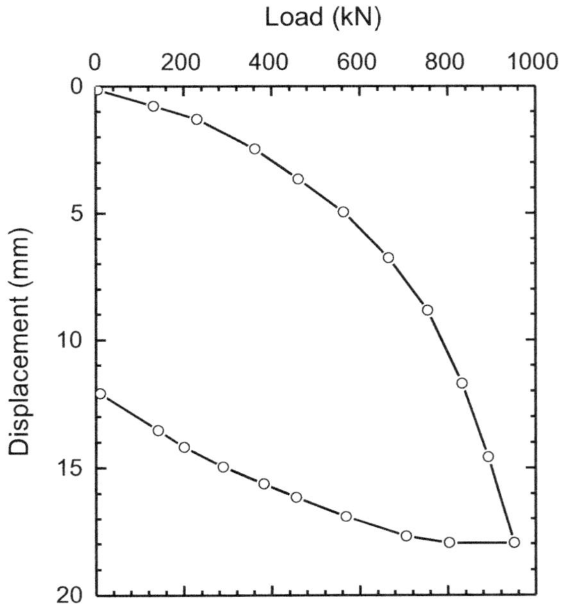

Detail	Description
Pile type/material	Open-ended steel pile
Length, L (m)	21.5
Outer diameter, D (mm)	508
Wall thickness, t (mm)	6.3
Installation method	Driven
Set up time, days	43
Loading mode	Tension
$Q_{max\text{-}measured}$ (kN)	980
Q_m (kN)	980
Q_s (kN)	980
Q_b (kN)	—
API Q_c (kN)	978
Q_c/Q_m	1.00
UWA Q_c (kN)	511
Q_c/Q_m	0.52
ICP Q_c (kN)	489
Q_c/Q_m	0.50
Fugro Q_c (kN)	261
Q_c/Q_m	0.27
NGI Q_c (kN)	440
Q_c/Q_m	0.45

Pile ID: Larvik L2-1

Load–displacement data

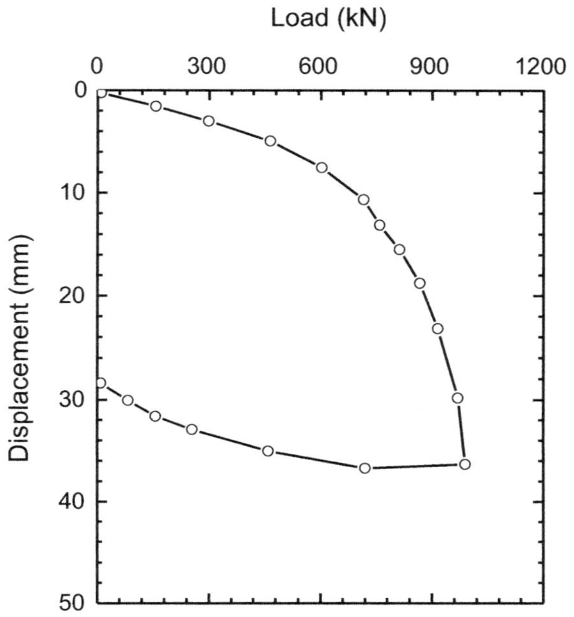

Detail	Description
Pile type/material	Open-ended steel pile
Length, L (m)	21.5
Outer diameter, D (mm)	508
Wall thickness, t (mm)	6.3
Installation method	Driven
Set up time, days	135
Loading mode	Tension
$Q_{max\text{-}measured}$ (kN)	990
Q_m (kN)	990
Q_s (kN)	990
Q_b (kN)	—
API Q_c (kN)	978
Q_c/Q_m	0.99
UWA Q_c (kN)	511
Q_c/Q_m	0.52
ICP Q_c (kN)	489
Q_c/Q_m	0.49
Fugro Q_c (kN)	261
Q_c/Q_m	0.26
NGI Q_c (kN)	440
Q_c/Q_m	0.44

Pile ID: Larvik L3-1

Load–displacement data

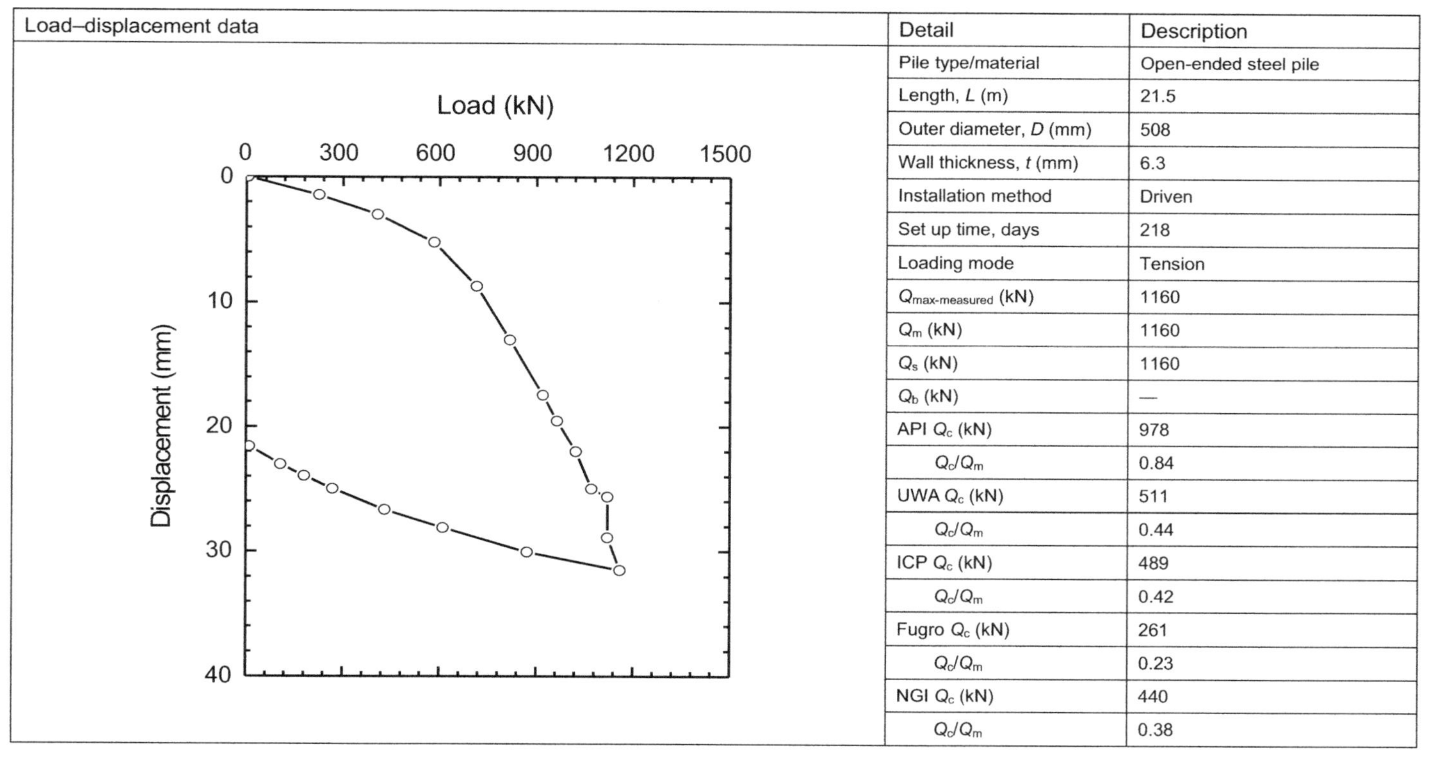

Detail	Description
Pile type/material	Open-ended steel pile
Length, L (m)	21.5
Outer diameter, D (mm)	508
Wall thickness, t (mm)	6.3
Installation method	Driven
Set up time, days	218
Loading mode	Tension
$Q_{max\text{-}measured}$ (kN)	1160
Q_m (kN)	1160
Q_s (kN)	1160
Q_b (kN)	—
API Q_c (kN)	978
Q_c/Q_m	0.84
UWA Q_c (kN)	511
Q_c/Q_m	0.44
ICP Q_c (kN)	489
Q_c/Q_m	0.42
Fugro Q_c (kN)	261
Q_c/Q_m	0.23
NGI Q_c (kN)	440
Q_c/Q_m	0.38

Site ID No. 24

Pile ID: Larvik L4-1

Load–displacement data

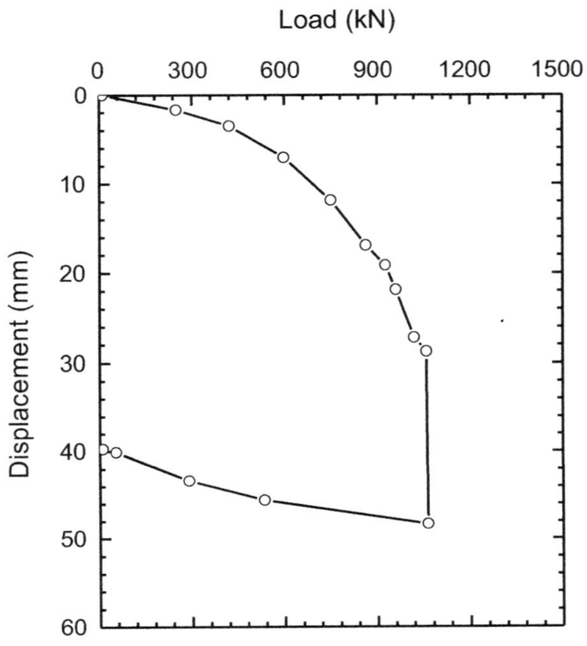

Detail	Description
Pile type/material	Open-ended steel pile
Length, L (m)	21.5
Outer diameter, D (mm)	508
Wall thickness, t (mm)	6.3
Installation method	Driven
Set up time, days	365
Loading mode	Tension
$Q_{max\text{-}measured}$ (kN)	1065
Q_m (kN)	1065
Q_s (kN)	1065
Q_b (kN)	—
API Q_c (kN)	978
Q_c/Q_m	0.92
UWA Q_c (kN)	511
Q_c/Q_m	0.48
ICP Q_c (kN)	489
Q_c/Q_m	0.46
Fugro Q_c (kN)	261
Q_c/Q_m	0.25
NGI Q_c (kN)	440
Q_c/Q_m	0.41

Pile ID: Larvik L5-1

Load–displacement data

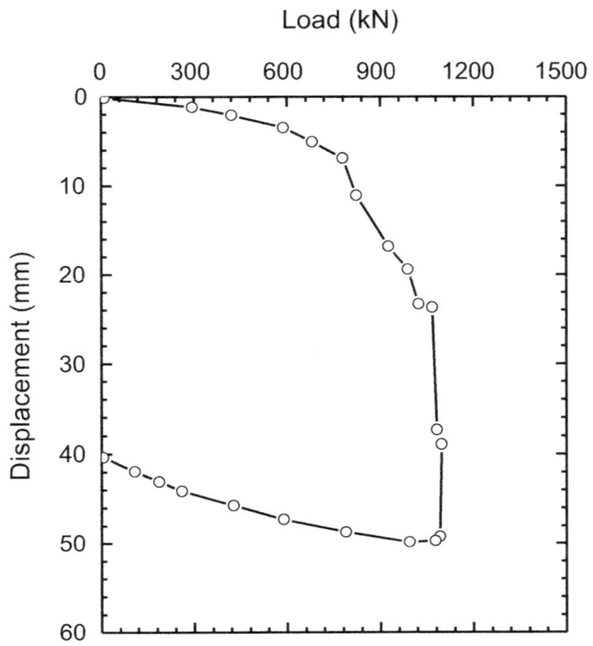

Detail	Description
Pile type/material	Open-ended steel pile
Length, L (m)	21.5
Outer diameter, D (mm)	508
Wall thickness, t (mm)	6.3
Installation method	Driven
Set up time, days	730
Loading mode	Tension
$Q_{max\text{-}measured}$ (kN)	1080
Q_m (kN)	1080
Q_s (kN)	1080
Q_b (kN)	—
API Q_c (kN)	978
Q_c/Q_m	0.91
UWA Q_c (kN)	511
Q_c/Q_m	0.47
ICP Q_c (kN)	489
Q_c/Q_m	0.45
Fugro Q_c (kN)	261
Q_c/Q_m	0.24
NGI Q_c (kN)	440
Q_c/Q_m	0.41

Pile ID: Larvik L6-1

Load–displacement data

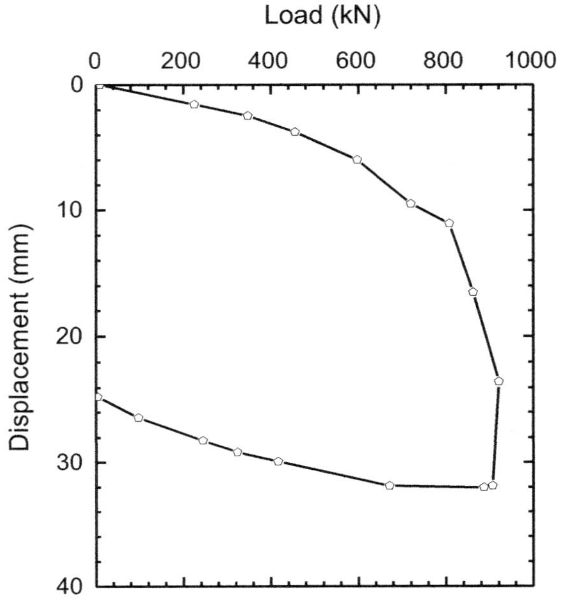

Detail	Description
Pile type/material	Open-ended steel pile
Length, L (m)	21.5
Outer diameter, D (mm)	508
Wall thickness, t (mm)	6.3
Installation method	Driven
Set up time, days	730
Loading mode	Tension
$Q_{max\text{-}measured}$ (kN)	900
Q_m (kN)	900
Q_s (kN)	900
Q_b (kN)	—
API Q_c (kN)	978
Q_c/Q_m	1.09
UWA Q_c (kN)	511
Q_c/Q_m	0.57
ICP Q_c (kN)	489
Q_c/Q_m	0.54
Fugro Q_c (kN)	261
Q_c/Q_m	0.29
NGI Q_c (kN)	440
Q_c/Q_m	0.49

Pile ID: Larvik L7-1

Load–displacement data

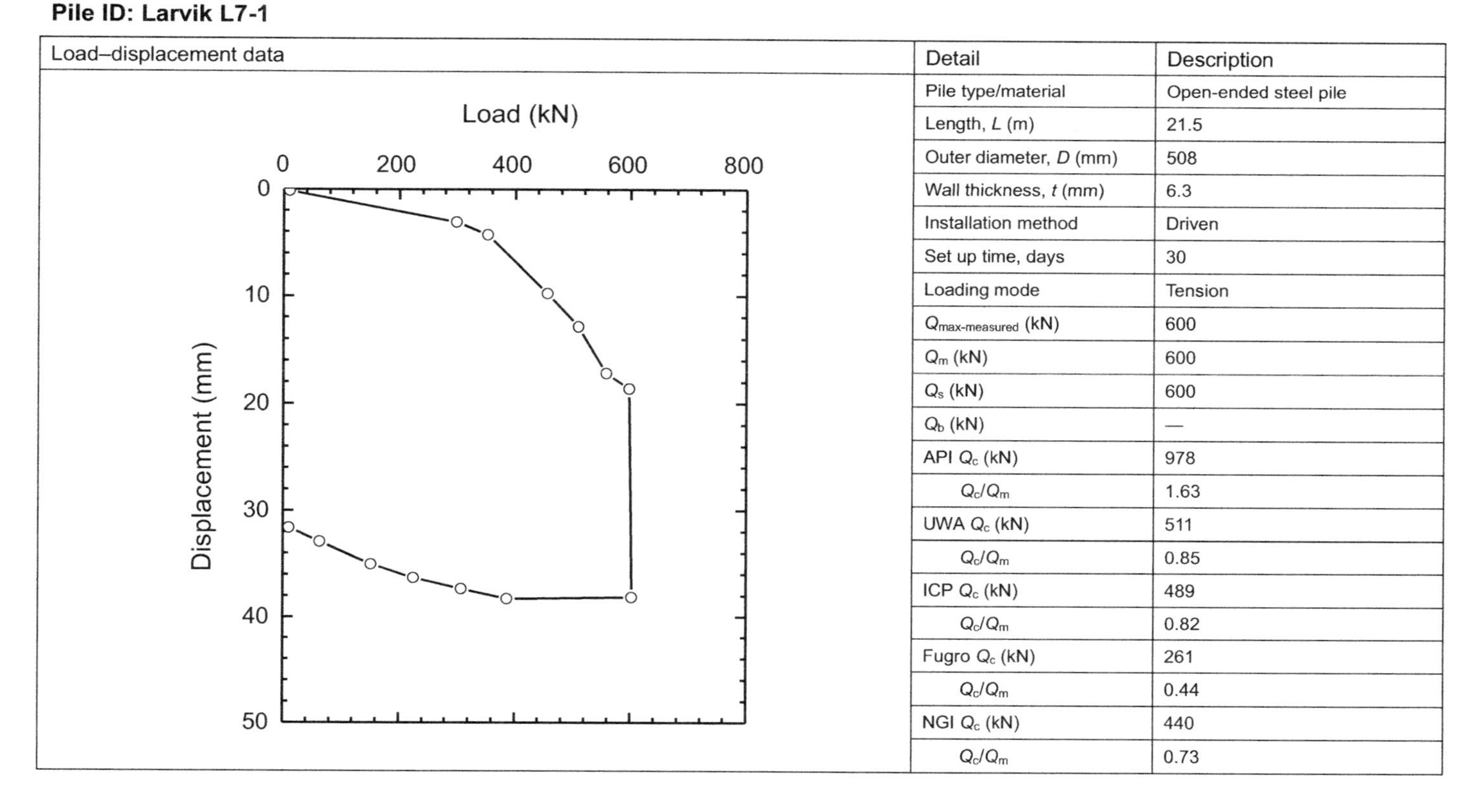

Detail	Description
Pile type/material	Open-ended steel pile
Length, L (m)	21.5
Outer diameter, D (mm)	508
Wall thickness, t (mm)	6.3
Installation method	Driven
Set up time, days	30
Loading mode	Tension
$Q_{max\text{-}measured}$ (kN)	600
Q_m (kN)	600
Q_s (kN)	600
Q_b (kN)	—
API Q_c (kN)	978
Q_c/Q_m	1.63
UWA Q_c (kN)	511
Q_c/Q_m	0.85
ICP Q_c (kN)	489
Q_c/Q_m	0.82
Fugro Q_c (kN)	261
Q_c/Q_m	0.44
NGI Q_c (kN)	440
Q_c/Q_m	0.73

Site ID No. 25: Jackson Country, USA.

Ref.: Mayne and Elhakim (2002): Axial pile response evaluation by geophysical piezocone tests. Proceedings of the 9th International Conference on Piling and Deep Foundations, DFI, Nice, Presses de l'ecole nationale des Ponts et chausses, 543–550.

Cone penetrometer data	Detail	Description
q_c (MPa) 0 5 10 15 20; Depth (m) 0 2 4 6 8 10 12 14 16 18 20	Site name and location	Jackson Country Electrical power Facility
	Soil type (s)	Silty sand and sandy silt
	Water table depth (m)	12
	Pile type (s)	Closed-ended steel pile
	Type of cone penetrometer testing	CPTU with F_r and pore pressure
	Number of pile load tests	1
	Comments	Two piles here, Pile JCEPF 1 is not accepted as its capacity is only half of JCEPF 2, but both with the same length and site conditions Interface friction angle and soil unit weight apply default value.

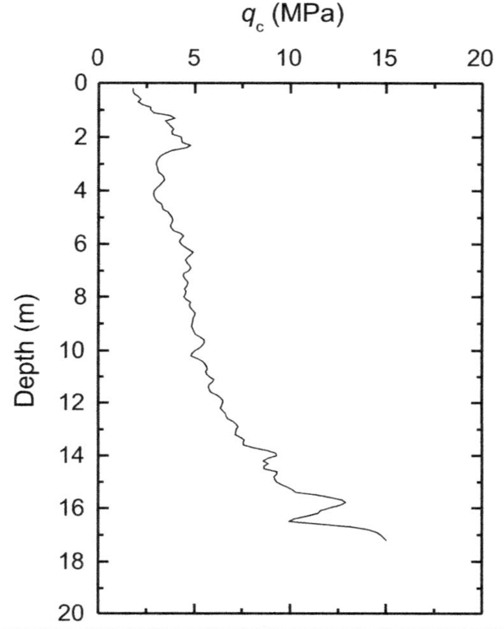

Pile ID: JCEPF 2

Load–displacement data

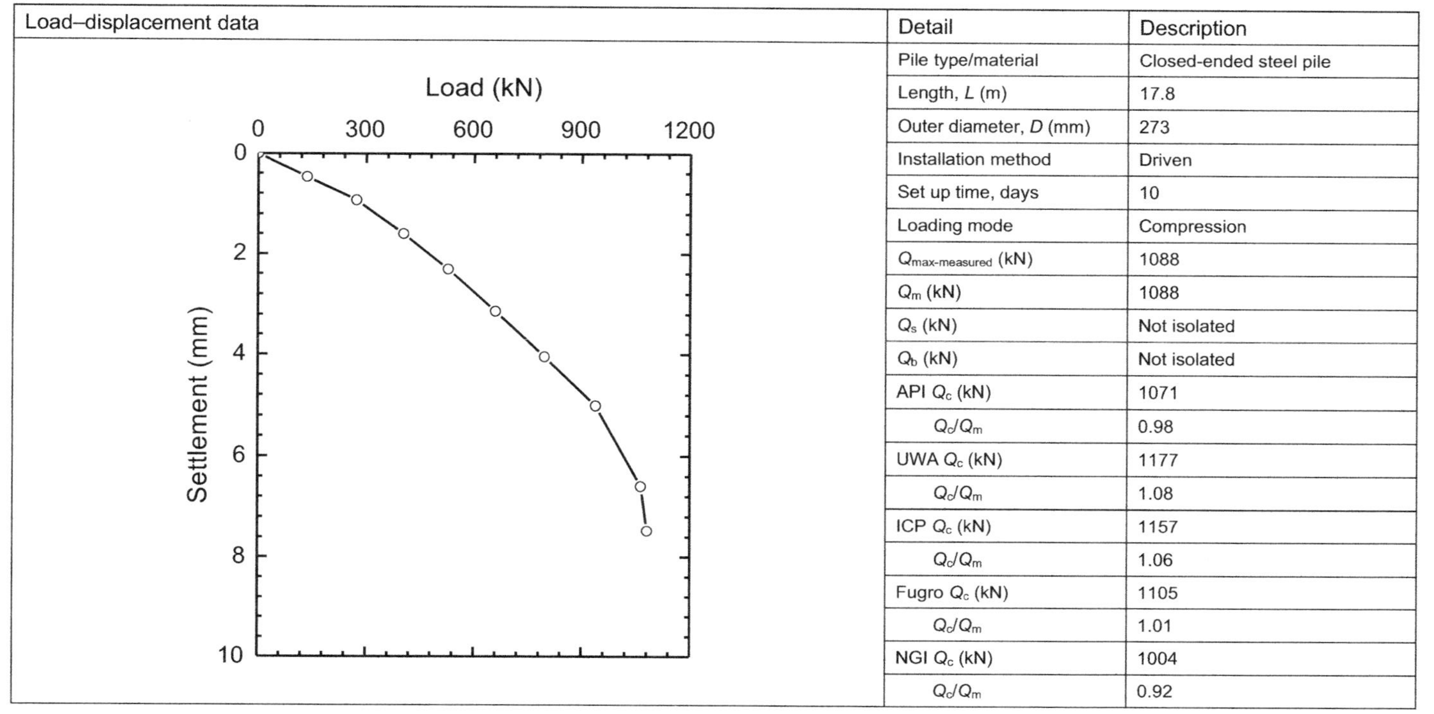

Detail	Description
Pile type/material	Closed-ended steel pile
Length, L (m)	17.8
Outer diameter, D (mm)	273
Installation method	Driven
Set up time, days	10
Loading mode	Compression
$Q_{max\text{-}measured}$ (kN)	1088
Q_m (kN)	1088
Q_s (kN)	Not isolated
Q_b (kN)	Not isolated
API Q_c (kN)	1071
Q_c/Q_m	0.98
UWA Q_c (kN)	1177
Q_c/Q_m	1.08
ICP Q_c (kN)	1157
Q_c/Q_m	1.06
Fugro Q_c (kN)	1105
Q_c/Q_m	1.01
NGI Q_c (kN)	1004
Q_c/Q_m	0.92

Site ID No. 26: Lafayette Bridge, USA.

Ref.: Komurka et al. (2010): Pile test program report: Lafayette Bridge Replacement. Report No. 09019 submitted to Minnesota Department of Transportation, by Wagner Komurka Geotechnical Group, Inc., Cedarburg, WI: 844p.

Cone penetrometer data

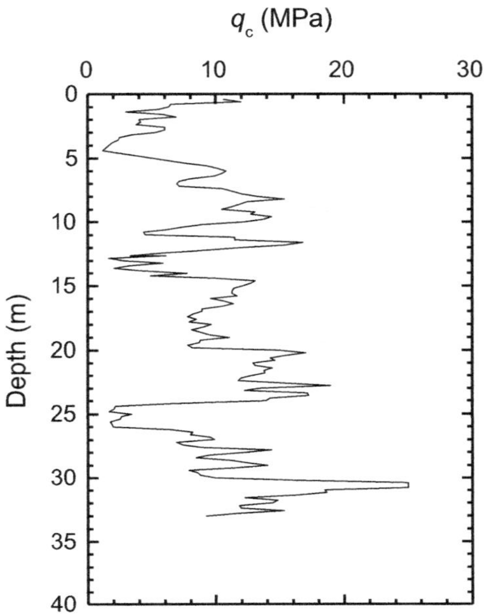

q_c (MPa)
0 10 20 30
Depth (m)
0 5 10 15 20 25 30 35 40

Detail	Description
Site name and location	Trunk Hwy 52 Lafayette Bridge
Soil type (s)	Silty sand and sandy silt
Water table depth (m)	5
Pile type (s)	Closed-ended steel pile
Type of cone penetrometer testing	CPTU with F_r and pore pressure
Number of pile load tests	1
Comments	Two piles here but only one is accepted, as CPT q_c values are not avaibale over whole length of Pile TH52 PAT2. Interface friction angle and soil unit weight apply default value.

Pile ID: TH52 MAT2

Load–displacement data

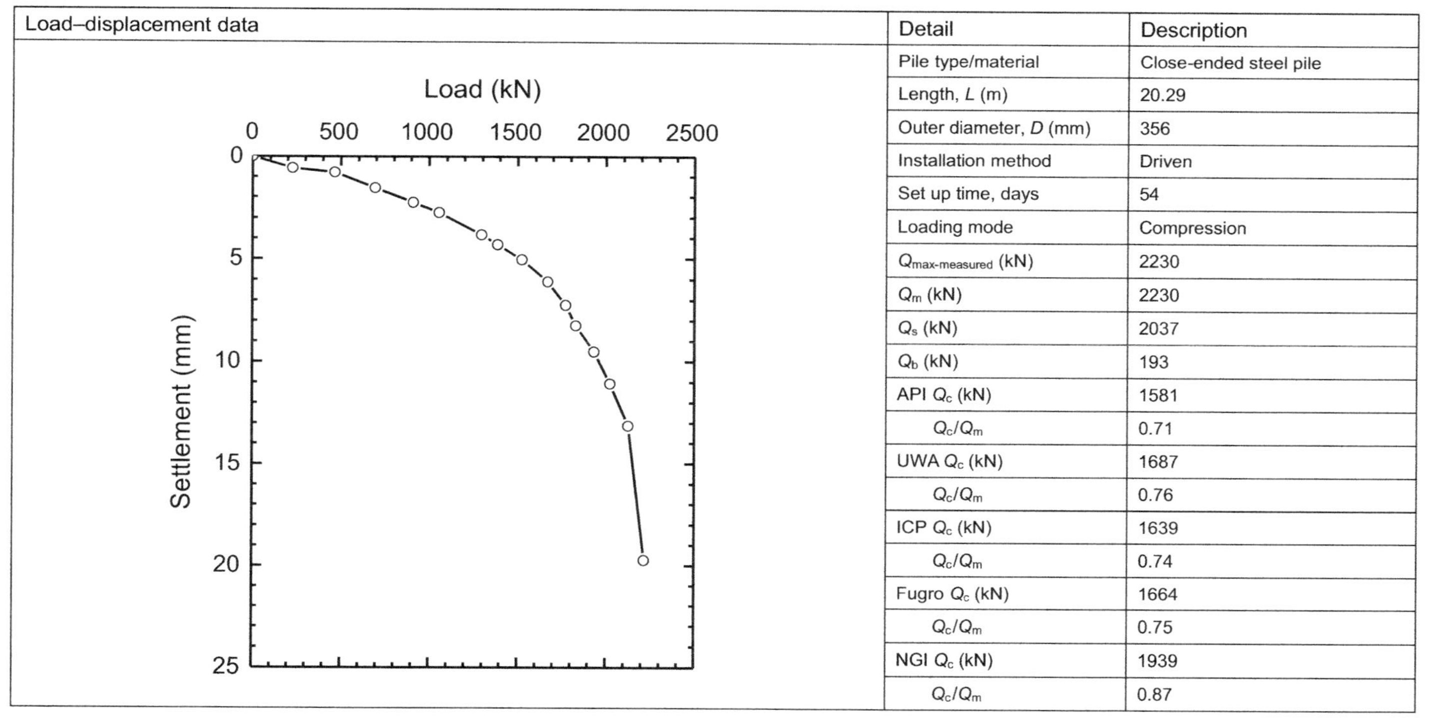

Detail	Description
Pile type/material	Close-ended steel pile
Length, L (m)	20.29
Outer diameter, D (mm)	356
Installation method	Driven
Set up time, days	54
Loading mode	Compression
$Q_{max\text{-}measured}$ (kN)	2230
Q_m (kN)	2230
Q_s (kN)	2037
Q_b (kN)	193
API Q_c (kN)	1581
Q_c/Q_m	0.71
UWA Q_c (kN)	1687
Q_c/Q_m	0.76
ICP Q_c (kN)	1639
Q_c/Q_m	0.74
Fugro Q_c (kN)	1664
Q_c/Q_m	0.75
NGI Q_c (kN)	1939
Q_c/Q_m	0.87

Site ID No. 27: Ogechee River, USA.

Ref.: Vesic (1970): Tests on instrumented piles, Ogeechee River site. J. Soil Mech. and Found. Div., 96(2), 561–584.

Cone penetrometer data

Detail	Description
Site name and location	Ogechee River
Soil type (s)	Medium dense, brown-gray fine sand, slightly silty
Water table depth (m)	1.5
Pile type (s)	One square concrete pile and five closed-ended steel piles
Type of cone penetrometer testing	CPT
Number of pile load tests	6
Comments	All piles were instrumented and the set up time was less than two days. Interface friction angle estimated with PSD, soil unit weight estimated by nuclear density test. We simply use the data given in the table.

Pile ID: H-2

Load–displacement data

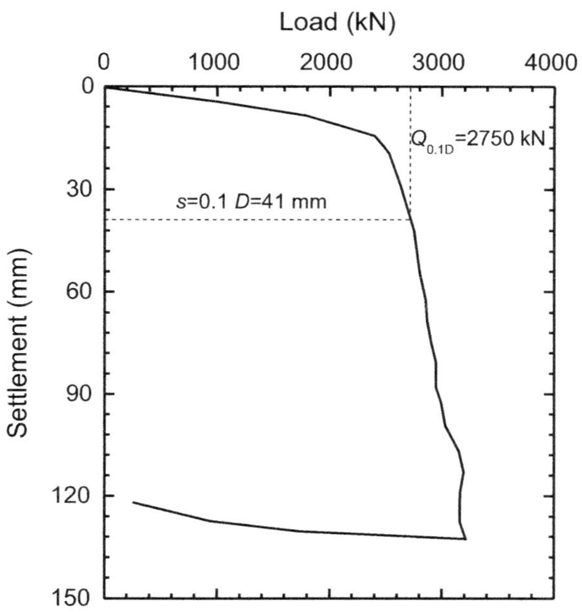

Detail	Description
Pile type/material	Square concrete pile
Length, L (m)	15.2
Outer width, B (mm)	406
Installation method	Driven
Set up time, days	0.5
Loading mode	Compression
$Q_{max\text{-}measured}$ (kN)	3160
Q_m (kN)	2750
Q_s (kN)	1189
Q_b (kN)	1561
API Q_c (kN)	2116
Q_c/Q_m	0.77
UWA Q_c (kN)	2836
Q_c/Q_m	1.03
ICP Q_c (kN)	2229
Q_c/Q_m	0.81
Fugro Q_c (kN)	3200
Q_c/Q_m	1.16
NGI Q_c (kN)	3499
Q_c/Q_m	1.27

Site ID No. 27

Pile ID: H-12

Load–displacement data

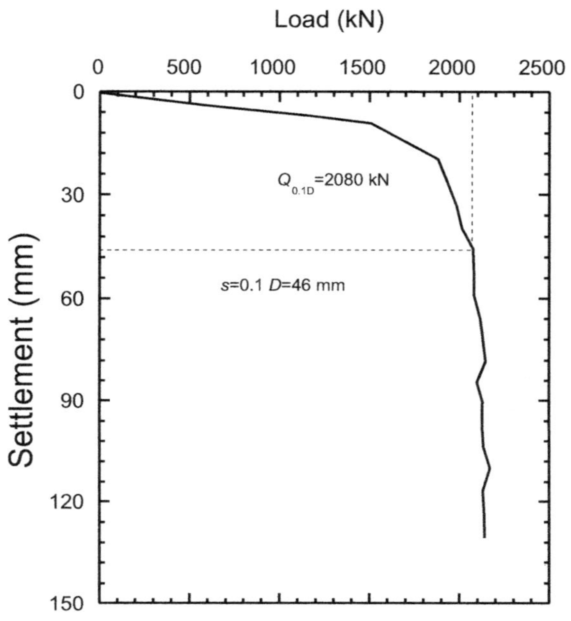

Detail	Description
Pile type/material	Closed-ended steel pipe pile
Length, L (m)	6.1
Outer diameter, D (mm)	457
Installation method	Driven
Set up time, days	0.5
Loading mode	Compression
$Q_{max\text{-}measured}$ (kN)	2140
Q_m (kN)	2080
Q_s (kN)	529
Q_b (kN)	1551
API Q_c (kN)	424
Q_c/Q_m	0.20
UWA Q_c (kN)	881
Q_c/Q_m	0.42
ICP Q_c (kN)	1317
Q_c/Q_m	0.63
Fugro Q_c (kN)	2270
Q_c/Q_m	1.09
NGI Q_c (kN)	1637
Q_c/Q_m	0.79

Pile ID: H-13

Load–displacement data

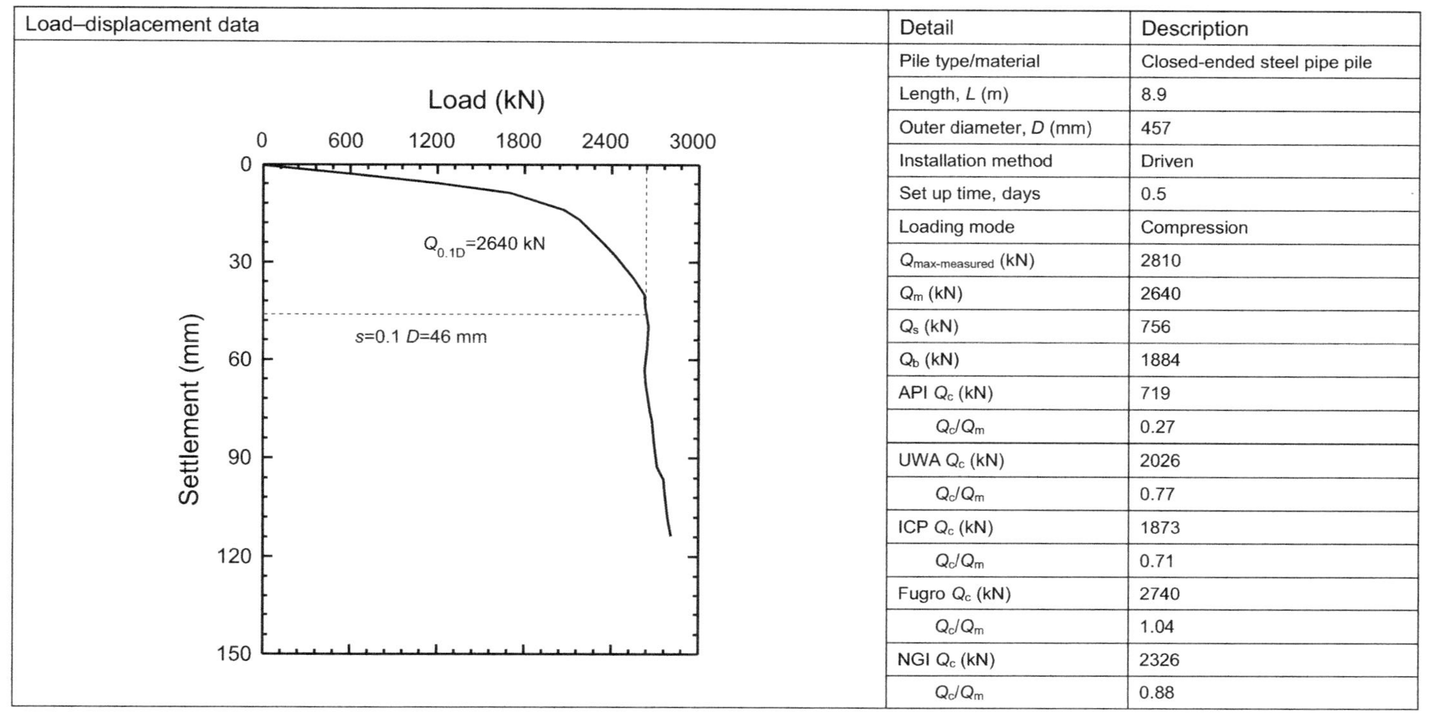

Detail	Description
Pile type/material	Closed-ended steel pipe pile
Length, L (m)	8.9
Outer diameter, D (mm)	457
Installation method	Driven
Set up time, days	0.5
Loading mode	Compression
$Q_{max\text{-}measured}$ (kN)	2810
Q_m (kN)	2640
Q_s (kN)	756
Q_b (kN)	1884
API Q_c (kN)	719
Q_c/Q_m	0.27
UWA Q_c (kN)	2026
Q_c/Q_m	0.77
ICP Q_c (kN)	1873
Q_c/Q_m	0.71
Fugro Q_c (kN)	2740
Q_c/Q_m	1.04
NGI Q_c (kN)	2326
Q_c/Q_m	0.88

Pile ID: H-14

Load–displacement data

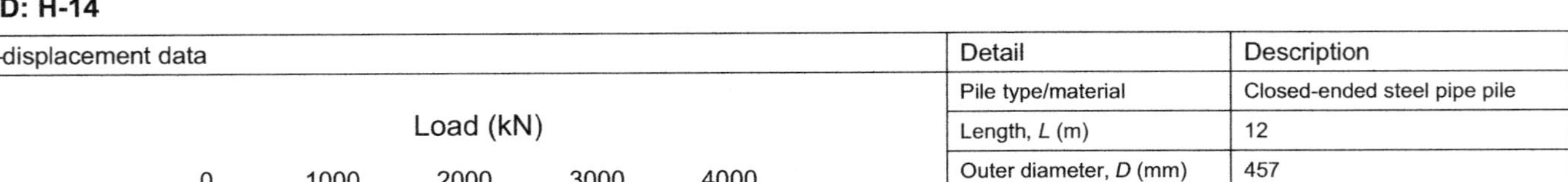

Detail	Description
Pile type/material	Closed-ended steel pipe pile
Length, L (m)	12
Outer diameter, D (mm)	457
Installation method	Driven
Set up time, days	0.5
Loading mode	Compression
$Q_{max\text{-}measured}$ (kN)	3560
Q_m (kN)	3210
Q_s (kN)	1243
Q_b (kN)	1967
API Q_c (kN)	1134
Q_c/Q_m	0.35
UWA Q_c (kN)	2248
Q_c/Q_m	0.70
ICP Q_c (kN)	2181
Q_c/Q_m	0.68
Fugro Q_c (kN)	3030
Q_c/Q_m	0.94
NGI Q_c (kN)	2507
Q_c/Q_m	0.78

Pile ID: H-15

Load–displacement data

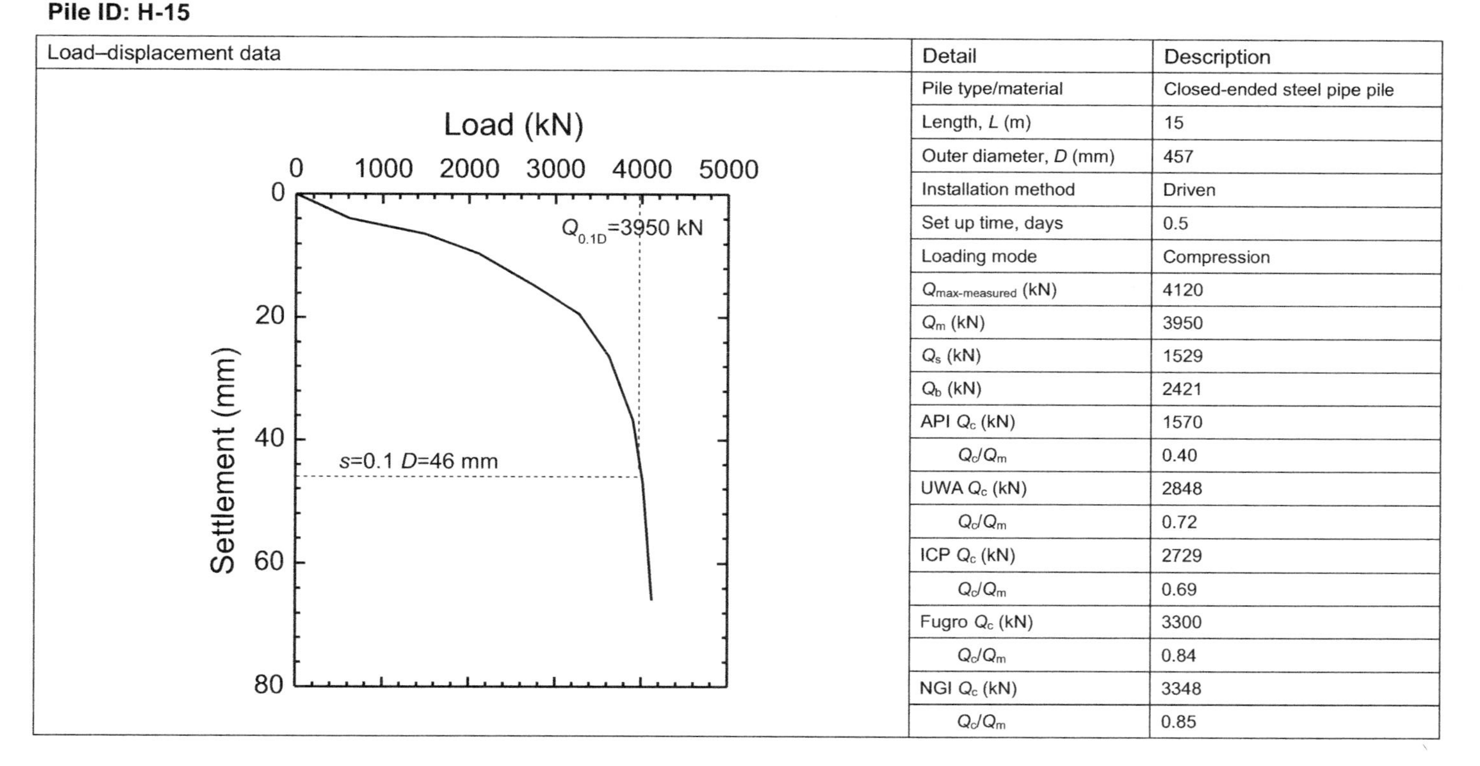

Detail	Description
Pile type/material	Closed-ended steel pipe pile
Length, L (m)	15
Outer diameter, D (mm)	457
Installation method	Driven
Set up time, days	0.5
Loading mode	Compression
$Q_{max\text{-}measured}$ (kN)	4120
Q_m (kN)	3950
Q_s (kN)	1529
Q_b (kN)	2421
API Q_c (kN)	1570
Q_c/Q_m	0.40
UWA Q_c (kN)	2848
Q_c/Q_m	0.72
ICP Q_c (kN)	2729
Q_c/Q_m	0.69
Fugro Q_c (kN)	3300
Q_c/Q_m	0.84
NGI Q_c (kN)	3348
Q_c/Q_m	0.85

Pile ID: H-16

Load–displacement data	Detail	Description
Tabulated results but no curves given	Pile type/material	Closed-ended steel pipe pile
	Length, L (m)	15
	Outer diameter, D (mm)	457
	Installation method	Driven
	Set up time, days	15
	Loading mode	tension
	$Q_{max\text{-}measured}$ (kN)	1540
	Q_m (kN)	1540
	Q_s (kN)	1540
	Q_b (kN)	—
	API Q_c (kN)	1005
	Q_c/Q_m	0.65
	UWA Q_c (kN)	1142
	Q_c/Q_m	0.74
	ICP Q_c (kN)	1153
	Q_c/Q_m	0.75
	Fugro Q_c (kN)	1120
	Q_c/Q_m	0.73
	NGI Q_c (kN)	1546
	Q_c/Q_m	1.00

Site ID No. 28: Drammen, Norway.

Ref.: Gregersen et al. (1973): Load tests on friction piles in loose sand. Proc. 8th Int. Conf. on Soil Mechanics and Foundation Engineering, Vol. 2.1, 109–117.

Cone penetrometer data	Detail	Description
q_c (MPa) 0 4 8 12; Depth (m) 0 5 10 15 20 25	Site name and location	A small island called Homen, Drammen
	Soil type (s)	loose sand
	Water table depth (m)	1.7
	Pile type (s)	Three closed-ended concrete pile
	Type of cone penetrometer testing	CPT
	Number of pile load tests	10
	Comments	Three piles altogether, tested in compression and tension after driving to various depths. Set up times were probably short. Only first tests from any given depth are accepted Interface friction angle applies default value. Soil unit weight was described in the original paper.

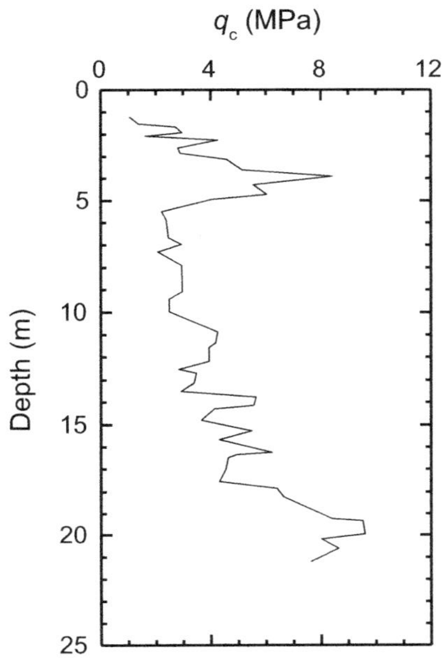

Pile ID: A

Load–displacement data

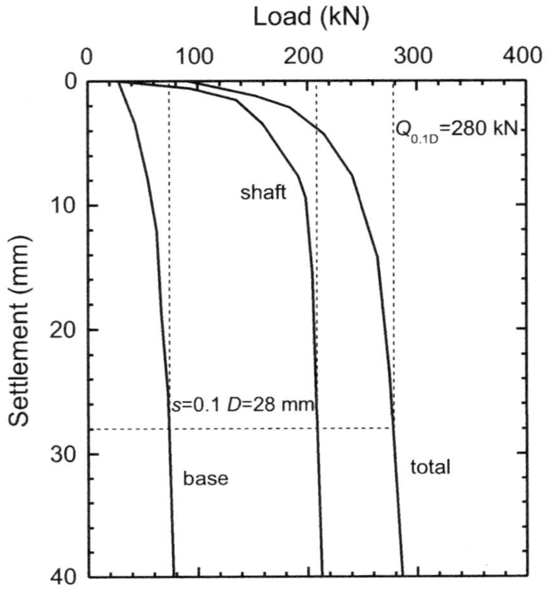

Detail	Description
Pile type/material	Closed-ended concrete pile
Length, L (m)	8
Outer diameter, D (mm)	280
Installation method	Driven
Set up time, days	—
Loading mode	Compression
$Q_{max\text{-}measured}$ (kN)	290
Q_m (kN)	280
Q_s (kN)	210
Q_b (kN)	70
API Q_c (kN)	184
Q_c/Q_m	0.66
UWA Q_c (kN)	255
Q_c/Q_m	0.91
ICP Q_c (kN)	237
Q_c/Q_m	0.85
Fugro Q_c (kN)	390
Q_c/Q_m	1.39
NGI Q_c (kN)	261
Q_c/Q_m	0.93

Pile ID: D/A

Load–displacement data

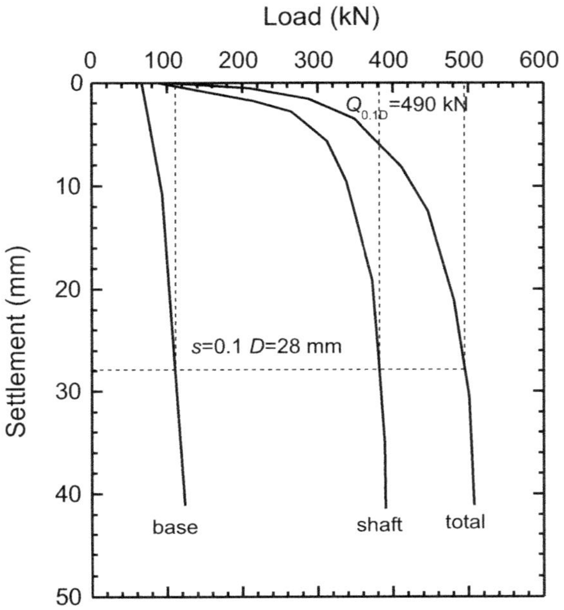

Detail	Description
Pile type/material	Closed-ended concrete pile
Length, L (m)	16
Outer diameter, D (mm)	280
Installation method	Driven
Set up time, days	—
Loading mode	Compression
$Q_{max\text{-}measured}$ (kN)	500
Q_m (kN)	490
Q_s (kN)	380
Q_b (kN)	110
API Q_c (kN)	534
Q_c/Q_m	1.09
UWA Q_c (kN)	494
Q_c/Q_m	1.01
ICP Q_c (kN)	490
Q_c/Q_m	1.00
Fugro Q_c (kN)	580
Q_c/Q_m	1.18
NGI Q_c (kN)	421
Q_c/Q_m	0.86

Site ID No. 28

Pile ID: E-7.5

Load–displacement data

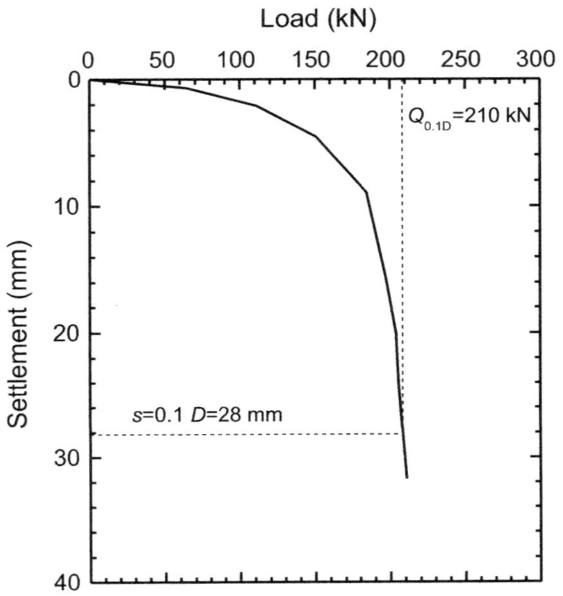

Detail	Description
Pile type/material	Closed-ended concrete pile
Length, L (m)	7.5
Outer diameter, D (mm)	280
Installation method	Driven
Set up time, days	—
Loading mode	Compression
$Q_{max\text{-}measured}$ (kN)	210
Q_m (kN)	210
Q_s (kN)	Not isolated
Q_b (kN)	Not isolated
API Q_c (kN)	150
Q_c/Q_m	0.71
UWA Q_c (kN)	237
Q_c/Q_m	1.13
ICP Q_c (kN)	221
Q_c/Q_m	1.05
Fugro Q_c (kN)	380
Q_c/Q_m	1.81
NGI Q_c (kN)	256
Q_c/Q_m	1.22

Pile ID: E-11.5

Load–displacement data

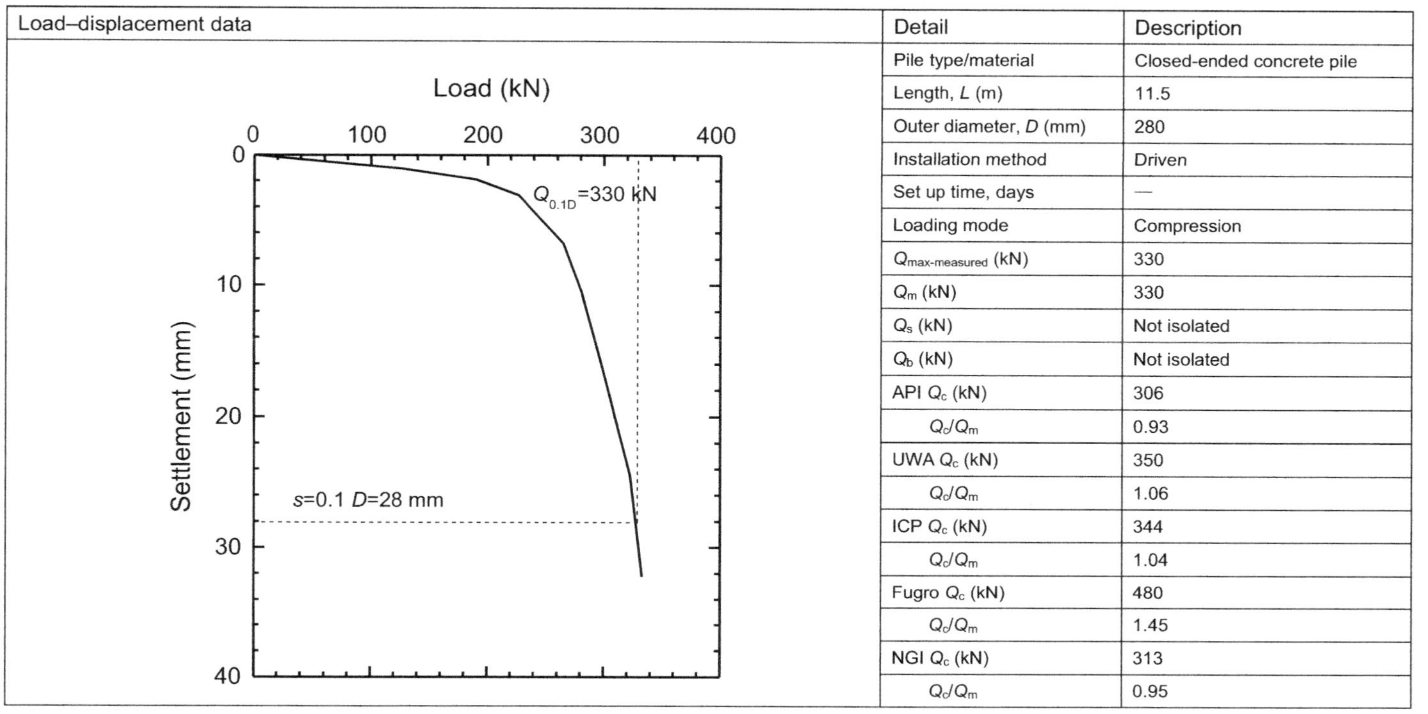

Detail	Description
Pile type/material	Closed-ended concrete pile
Length, L (m)	11.5
Outer diameter, D (mm)	280
Installation method	Driven
Set up time, days	—
Loading mode	Compression
$Q_{max\text{-}measured}$ (kN)	330
Q_m (kN)	330
Q_s (kN)	Not isolated
Q_b (kN)	Not isolated
API Q_c (kN)	306
Q_c/Q_m	0.93
UWA Q_c (kN)	350
Q_c/Q_m	1.06
ICP Q_c (kN)	344
Q_c/Q_m	1.04
Fugro Q_c (kN)	480
Q_c/Q_m	1.45
NGI Q_c (kN)	313
Q_c/Q_m	0.95

Pile ID: E-15.5

Load–displacement data

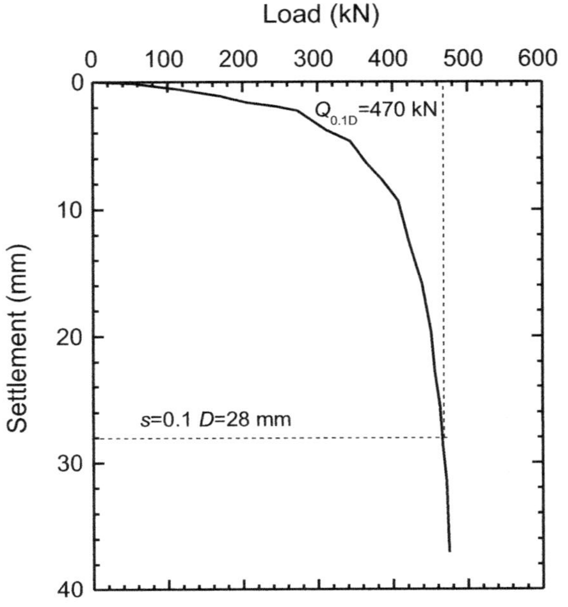
Load (kN)
0
100
200
300
400
500
600
0
10
20
30
40
Settlement (mm)
$Q_{0.1D}$=470 kN
s=0.1 D=28 mm

Detail	Description
Pile type/material	Closed-ended concrete pile
Length, L (m)	15.5
Outer diameter, D (mm)	280
Installation method	Driven
Set up time, days	—
Loading mode	Compression
$Q_{max\text{-}measured}$ (kN)	480
Q_m (kN)	470
Q_s (kN)	Not isolated
Q_b (kN)	Not isolated
API Q_c (kN)	505
Q_c/Q_m	1.05
UWA Q_c (kN)	467
Q_c/Q_m	0.99
ICP Q_c (kN)	472
Q_c/Q_m	1.01
Fugro Q_c (kN)	560
Q_c/Q_m	1.19
NGI Q_c (kN)	387
Q_c/Q_m	0.82

Pile ID: E-19.5

Load–displacement data

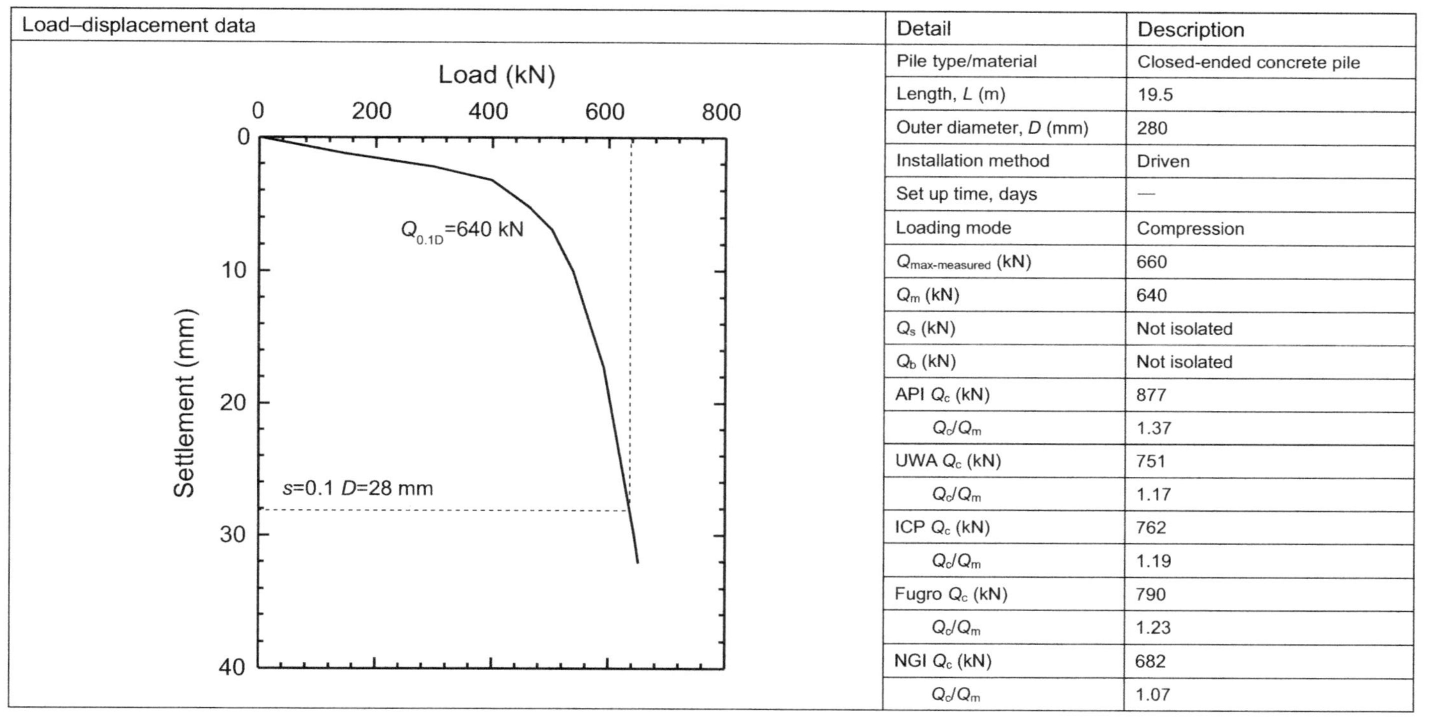

Detail	Description
Pile type/material	Closed-ended concrete pile
Length, L (m)	19.5
Outer diameter, D (mm)	280
Installation method	Driven
Set up time, days	—
Loading mode	Compression
$Q_{max\text{-}measured}$ (kN)	660
Q_m (kN)	640
Q_s (kN)	Not isolated
Q_b (kN)	Not isolated
API Q_c (kN)	877
Q_c/Q_m	1.37
UWA Q_c (kN)	751
Q_c/Q_m	1.17
ICP Q_c (kN)	762
Q_c/Q_m	1.19
Fugro Q_c (kN)	790
Q_c/Q_m	1.23
NGI Q_c (kN)	682
Q_c/Q_m	1.07

Site ID No. 28

Pile ID: E-23.5

Load–displacement data

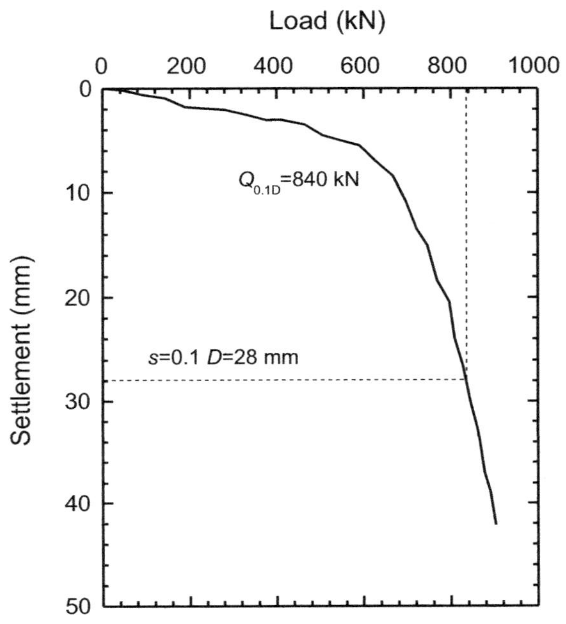

Detail	Description
Pile type/material	Closed-ended concrete pile
Length, L (m)	23.5
Outer diameter, D (mm)	280
Installation method	Driven
Set up time, days	—
Loading mode	Compression
$Q_{max\text{-}measured}$ (kN)	900
Q_m (kN)	840
Q_s (kN)	Not isolated
Q_b (kN)	Not isolated
API Q_c (kN)	1056
Q_c/Q_m	1.26
UWA Q_c (kN)	893
Q_c/Q_m	1.06
ICP Q_c (kN)	893
Q_c/Q_m	1.06
Fugro Q_c (kN)	850
Q_c/Q_m	1.01
NGI Q_c (kN)	827
Q_c/Q_m	0.98

Pile ID: A(T)

Load–displacement data

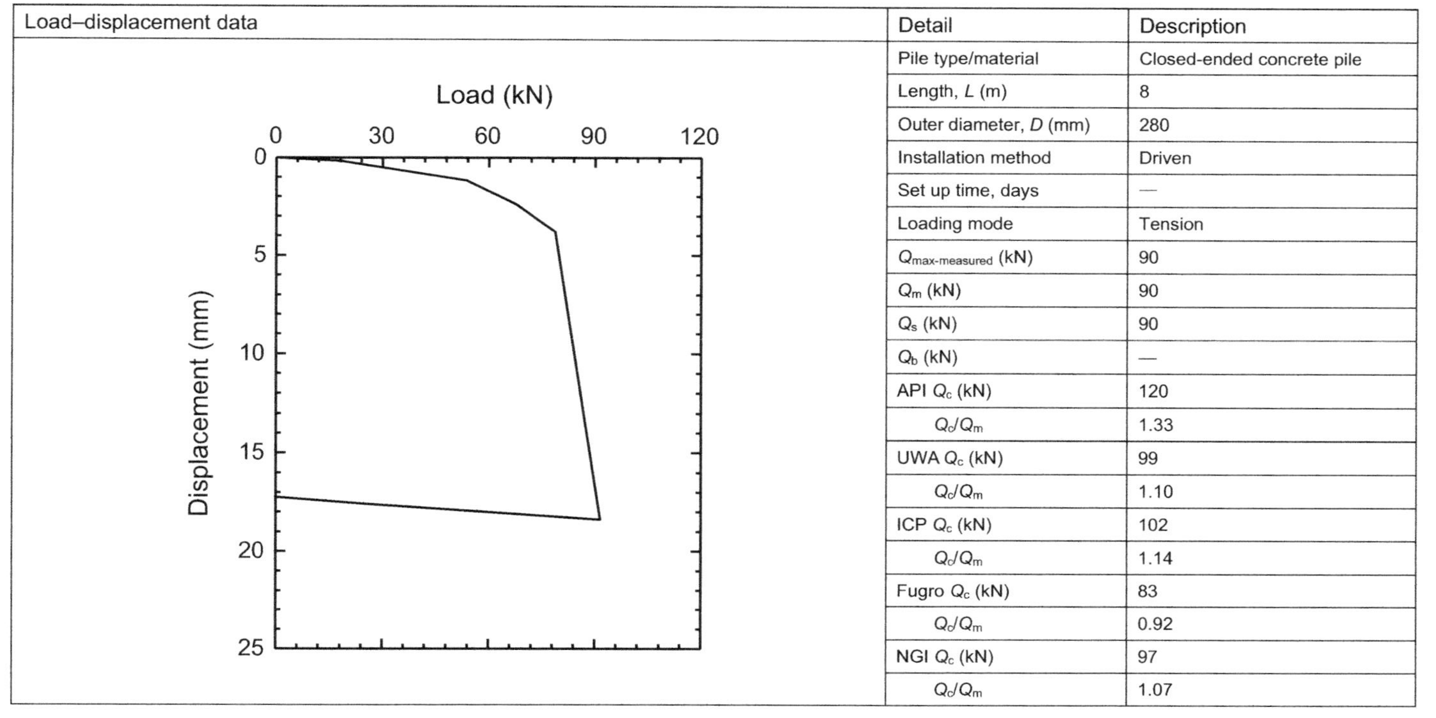

Detail	Description
Pile type/material	Closed-ended concrete pile
Length, L (m)	8
Outer diameter, D (mm)	280
Installation method	Driven
Set up time, days	—
Loading mode	Tension
$Q_{max\text{-}measured}$ (kN)	90
Q_m (kN)	90
Q_s (kN)	90
Q_b (kN)	—
API Q_c (kN)	120
Q_c/Q_m	1.33
UWA Q_c (kN)	99
Q_c/Q_m	1.10
ICP Q_c (kN)	102
Q_c/Q_m	1.14
Fugro Q_c (kN)	83
Q_c/Q_m	0.92
NGI Q_c (kN)	97
Q_c/Q_m	1.07

Pile ID: D/A(T)

Load–displacement data

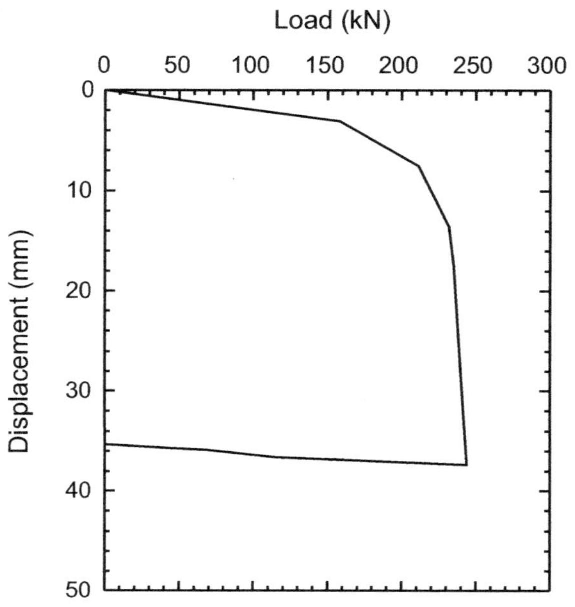

Load (kN)

Displacement (mm)

Detail	Description
Pile type/material	Closed-ended concrete pile
Length, L (m)	16
Outer diameter, D (mm)	280
Installation method	Driven
Set up time, days	—
Loading mode	Tension
$Q_{max\text{-}measured}$ (kN)	250
Q_m (kN)	250
Q_s (kN)	250
Q_b (kN)	—
API Q_c (kN)	410
Q_c/Q_m	1.64
UWA Q_c (kN)	202
Q_c/Q_m	0.81
ICP Q_c (kN)	230
Q_c/Q_m	0.92
Fugro Q_c (kN)	158
Q_c/Q_m	0.63
NGI Q_c (kN)	155
Q_c/Q_m	0.62

Pile ID: E- (T)

Load–displacement data

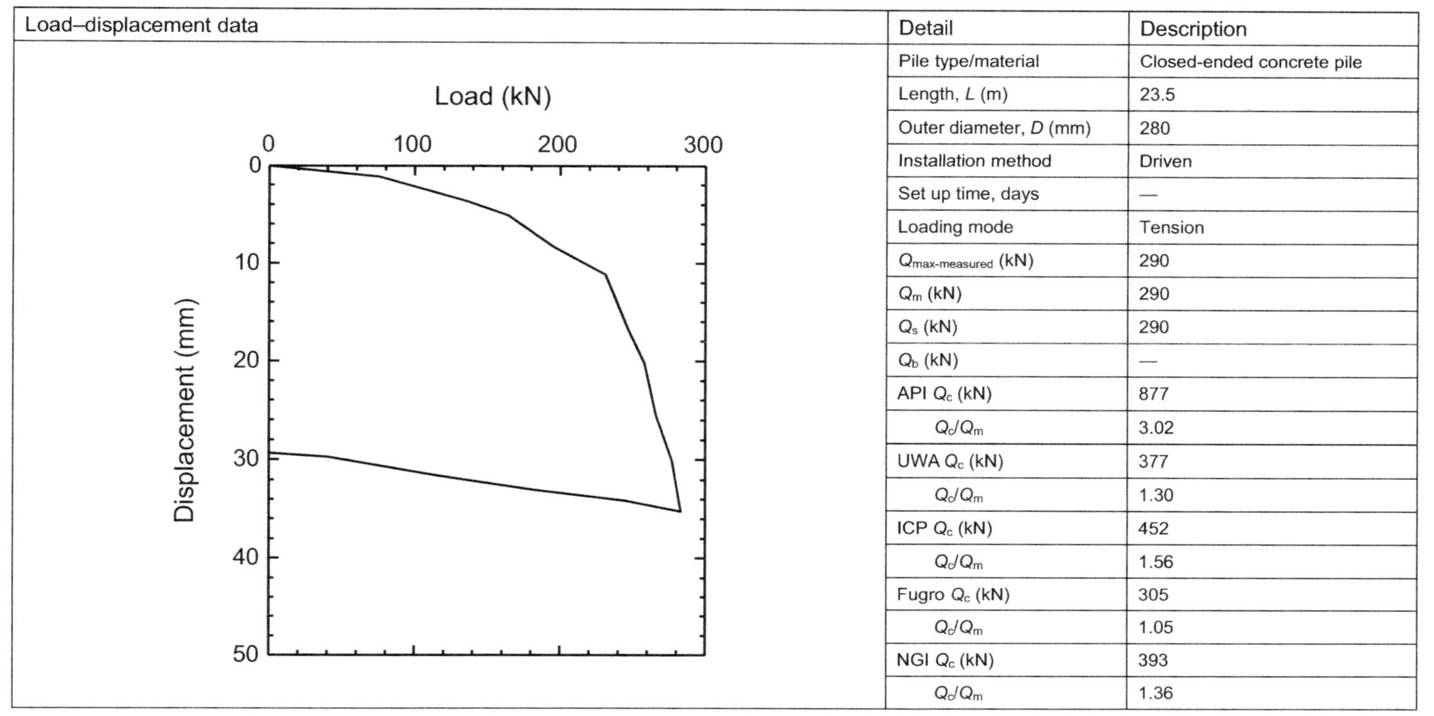

Detail	Description
Pile type/material	Closed-ended concrete pile
Length, L (m)	23.5
Outer diameter, D (mm)	280
Installation method	Driven
Set up time, days	—
Loading mode	Tension
$Q_{max\text{-}measured}$ (kN)	290
Q_m (kN)	290
Q_s (kN)	290
Q_b (kN)	—
API Q_c (kN)	877
Q_c/Q_m	3.02
UWA Q_c (kN)	377
Q_c/Q_m	1.30
ICP Q_c (kN)	452
Q_c/Q_m	1.56
Fugro Q_c (kN)	305
Q_c/Q_m	1.05
NGI Q_c (kN)	393
Q_c/Q_m	1.36

Site ID No. 29: Hoogzand, The Netherlands

Ref.: Beringen et al. (1979): Results of loading tests on driven piles in sand. Recent development in the design and construction of piles, ICE, London, 213–225.

Cone penetrometer data

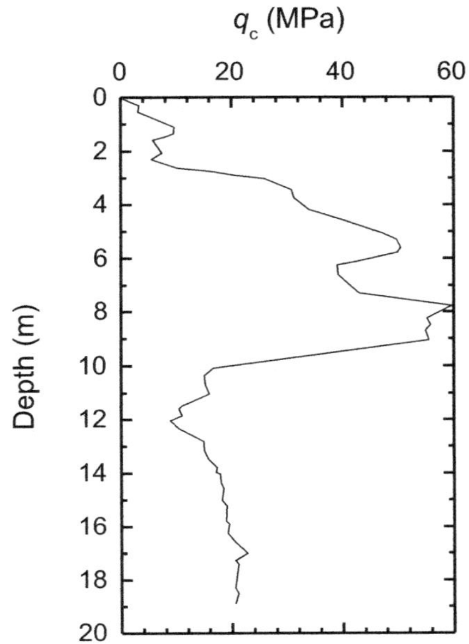

Detail	Description
Site name and location	Hoogzand, The Netherlands
Soil type (s)	Dense, locally very dense, Over-consolidated sand
Water table depth (m)	3.2
Pile type (s)	Two open-ended steel pipe piles and one closed-ended steel piles
Type of cone penetrometer testing	CPT
Number of pile load tests	6
Comments	Four piles reported but one pile was tested with a grouted plug and the results were not given. Piles under compression are highly instrumented. Pile capacity is also tabulated. Interface friction angle estimated by PSD. Soil unit weight use the value mentioned in original paper.

Pile ID: 1-C

Load–displacement data

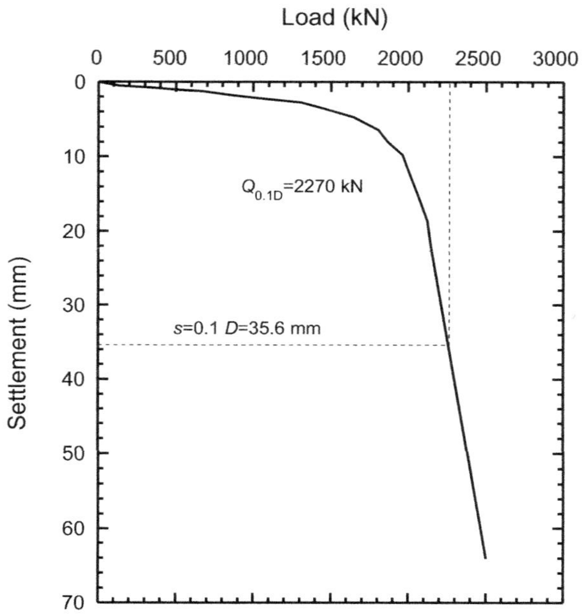

Load (kN)
0 500 1000 1500 2000 2500 3000
Settlement (mm)
0 10 20 30 40 50 60 70
$Q_{0.1D}$=2270 kN
s=0.1 D=35.6 mm

Detail	Description
Pile type/material	Open-ended steel pipe pile
Length, L (m)	7
Outer diameter, D (mm)	356
Wall thickness, t (mm)	16
Installation method	Driven
Set up time, days	37
Loading mode	Compression
$Q_{max\text{-}measured}$ (kN)	2500
Q_m (kN)	2270
Q_s (kN)	1310
Q_b (kN)	1130
API Q_c (kN)	707
Q_c/Q_m	0.31
UWA Q_c (kN)	1850
Q_c/Q_m	0.81
ICP Q_c (kN)	1512
Q_c/Q_m	0.67
Fugro Q_c (kN)	1710
Q_c/Q_m	0.75
NGI Q_c (kN)	1292
Q_c/Q_m	0.57

Site ID No. 29

Pile ID: 1-T

Load–displacement data

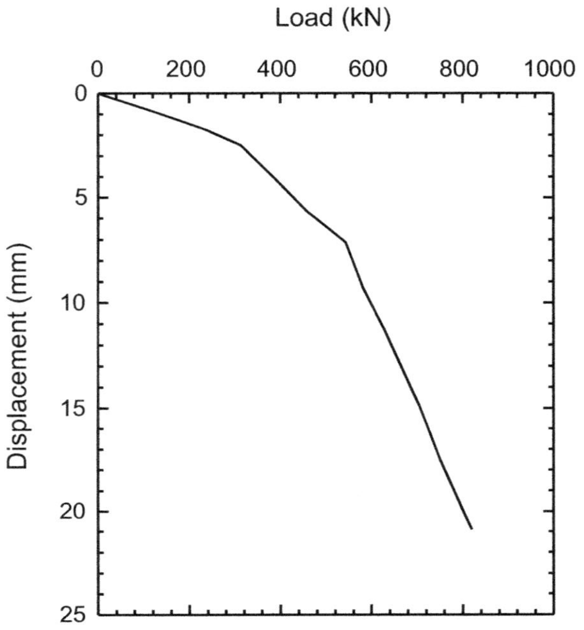

Detail	Description
Pile type/material	Open-ended steel pipe pile
Length, L (m)	7
Outer diameter, D (mm)	356
Wall thickness, t (mm)	16
Installation method	Driven
Set up time, days	37
Loading mode	Tension
$Q_{max\text{-}measured}$ (kN)	820
Q_m (kN)	820
Q_s (kN)	820
Q_b (kN)	—
API Q_c (kN)	234
Q_c/Q_m	0.29
UWA Q_c (kN)	683
Q_c/Q_m	0.83
ICP Q_c (kN)	562
Q_c/Q_m	0.69
Fugro Q_c (kN)	570
Q_c/Q_m	0.70
NGI Q_c (kN)	633
Q_c/Q_m	0.77

Pile ID: 3-C

Load–displacement data

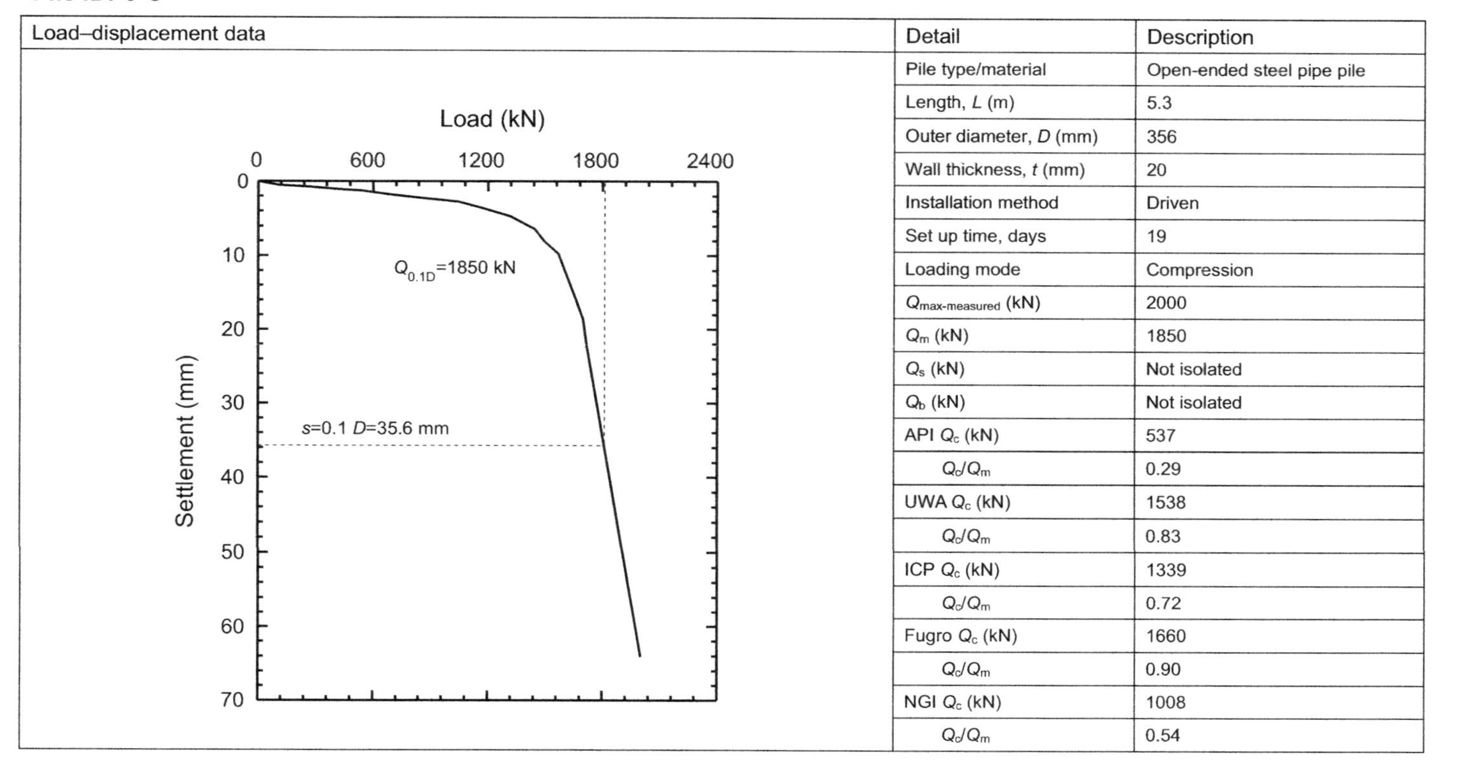

Detail	Description
Pile type/material	Open-ended steel pipe pile
Length, L (m)	5.3
Outer diameter, D (mm)	356
Wall thickness, t (mm)	20
Installation method	Driven
Set up time, days	19
Loading mode	Compression
$Q_{max\text{-}measured}$ (kN)	2000
Q_m (kN)	1850
Q_s (kN)	Not isolated
Q_b (kN)	Not isolated
API Q_c (kN)	537
Q_c/Q_m	0.29
UWA Q_c (kN)	1538
Q_c/Q_m	0.83
ICP Q_c (kN)	1339
Q_c/Q_m	0.72
Fugro Q_c (kN)	1660
Q_c/Q_m	0.90
NGI Q_c (kN)	1008
Q_c/Q_m	0.54

Site ID No. 29

Pile ID: 3-T

Load–displacement data

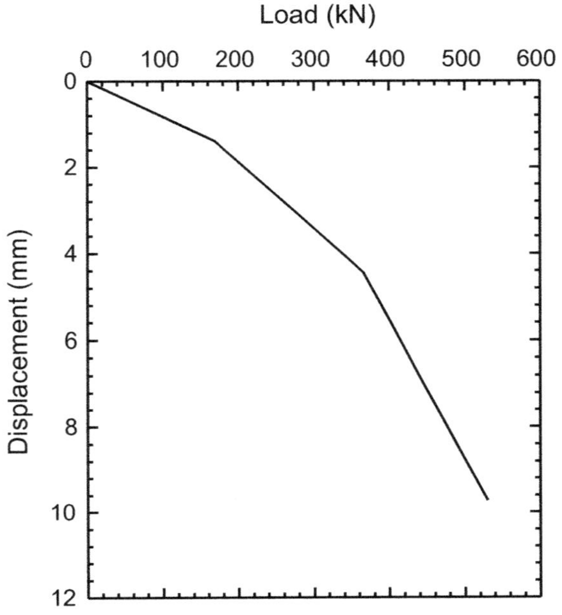

Detail	Description
Pile type/material	Open-ended steel pipe pile
Length, L (m)	5.3
Outer diameter, D (mm)	356
Wall thickness, t (mm)	20
Installation method	Driven
Set up time, days	19
Loading mode	Tension
$Q_{max\text{-}measured}$ (kN)	530
Q_m (kN)	530
Q_s (kN)	530
Q_b (kN)	—
API Q_c (kN)	140
Q_c/Q_m	0.26
UWA Q_c (kN)	476
Q_c/Q_m	0.90
ICP Q_c (kN)	376
Q_c/Q_m	0.71
Fugro Q_c (kN)	484
Q_c/Q_m	0.91
NGI Q_c (kN)	405
Q_c/Q_m	0.76

Pile ID: 2-C

Load–displacement data

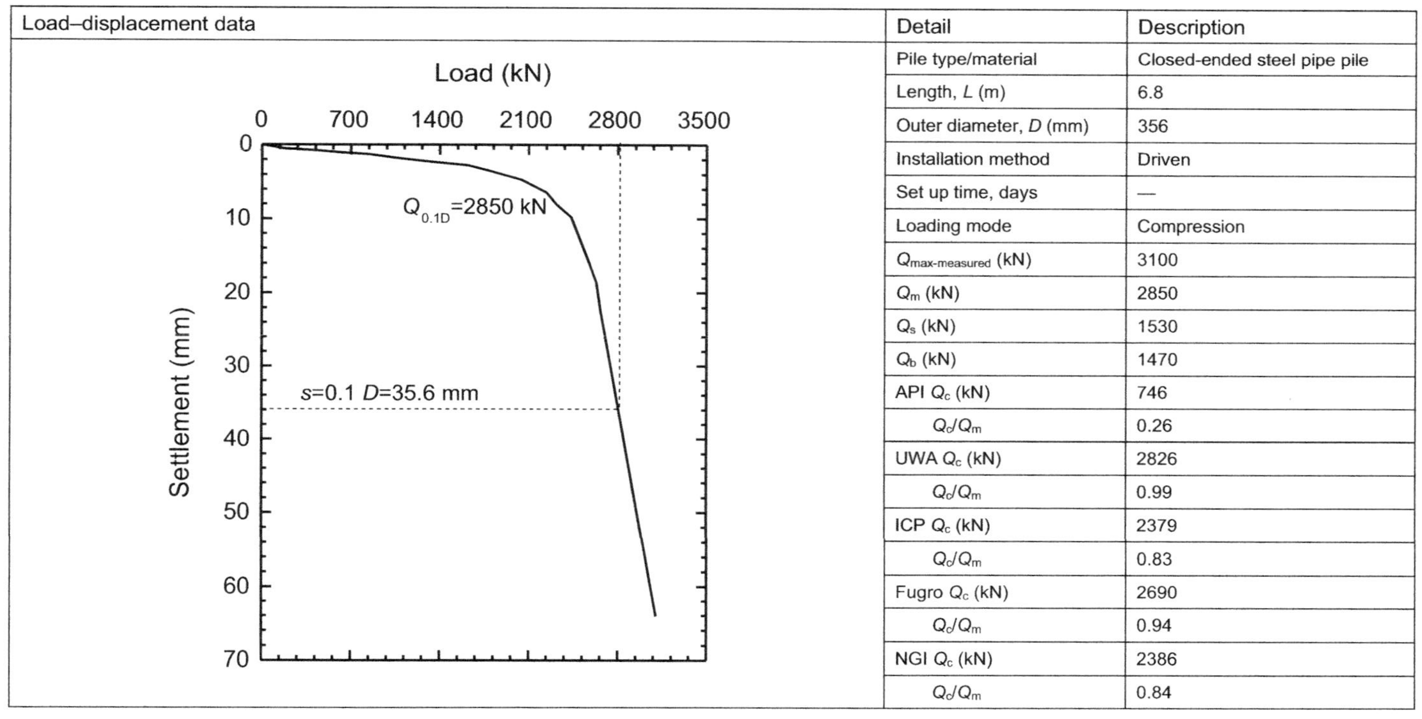

Detail	Description
Pile type/material	Closed-ended steel pipe pile
Length, L (m)	6.8
Outer diameter, D (mm)	356
Installation method	Driven
Set up time, days	—
Loading mode	Compression
$Q_{max\text{-}measured}$ (kN)	3100
Q_m (kN)	2850
Q_s (kN)	1530
Q_b (kN)	1470
API Q_c (kN)	746
Q_c/Q_m	0.26
UWA Q_c (kN)	2826
Q_c/Q_m	0.99
ICP Q_c (kN)	2379
Q_c/Q_m	0.83
Fugro Q_c (kN)	2690
Q_c/Q_m	0.94
NGI Q_c (kN)	2386
Q_c/Q_m	0.84

Site ID No. 29

Pile ID: 2-T

Load–displacement data

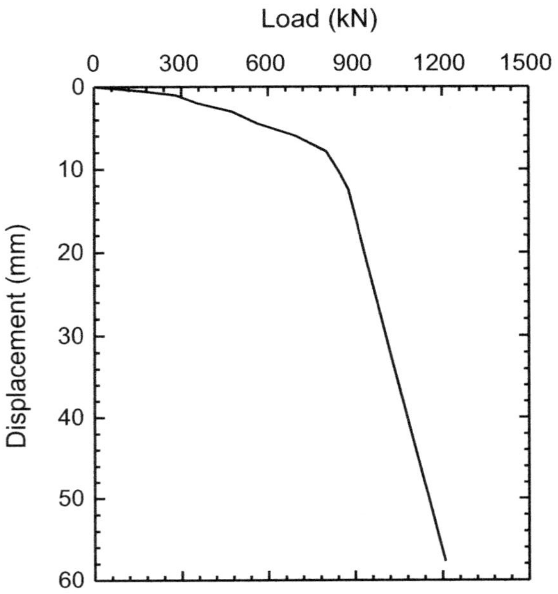

Detail	Description
Pile type/material	Closed-ended steel pipe pile
Length, L (m)	6.8
Outer diameter, D (mm)	356
Installation method	Driven
Set up time, days	—
Loading mode	Tension
$Q_{max\text{-}measured}$ (kN)	1210
Q_m (kN)	1210
Q_s (kN)	1210
Q_b (kN)	—
API Q_c (kN)	282
Q_c/Q_m	0.23
UWA Q_c (kN)	897
Q_c/Q_m	0.74
ICP Q_c (kN)	754
Q_c/Q_m	0.62
Fugro Q_c (kN)	964
Q_c/Q_m	0.80
NGI Q_c (kN)	991
Q_c/Q_m	0.82

Site ID No. 30: Hunter's Point, USA.

Ref.: Briaud et al. (1989a): Axially loaded 5 pile group and single pile in sand. Proc. 12th Int. Conf. on Soil Mechanics and Foundation Engineering, Balkema, Rotterdam, 1121–1124.

Cone penetrometer data

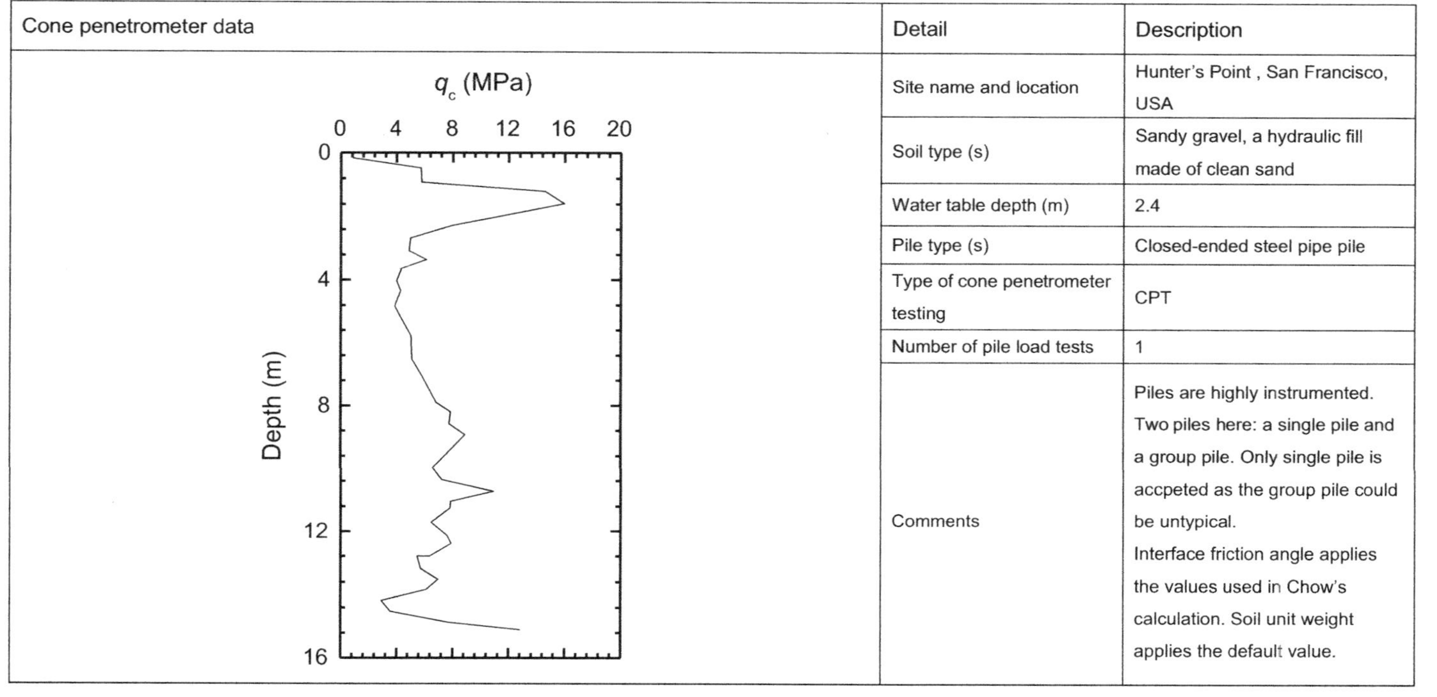

Detail	Description
Site name and location	Hunter's Point , San Francisco, USA
Soil type (s)	Sandy gravel, a hydraulic fill made of clean sand
Water table depth (m)	2.4
Pile type (s)	Closed-ended steel pipe pile
Type of cone penetrometer testing	CPT
Number of pile load tests	1
Comments	Piles are highly instrumented. Two piles here: a single pile and a group pile. Only single pile is accpeted as the group pile could be untypical. Interface friction angle applies the values used in Chow's calculation. Soil unit weight applies the default value.

Site ID No. 30

Pile ID: S

Load–displacement data

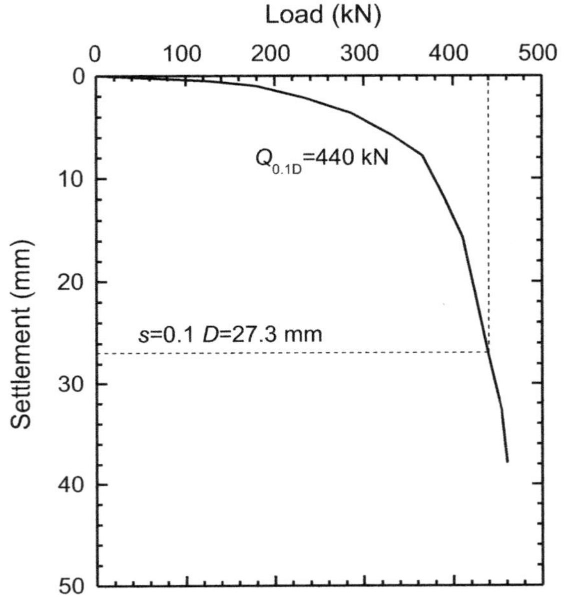

Detail	Description
Pile type/material	Closed-ended steel pipe pile
Length, L (m)	7.8
Outer diameter, D (mm)	273
Installation method	Driven
Set up time, days	24
Loading mode	Compression
$Q_{max\text{-}measured}$ (kN)	500
Q_m (kN)	440
Q_s (kN)	151
Q_b (kN)	289
API Q_c (kN)	497
Q_c/Q_m	1.13
UWA Q_c (kN)	493
Q_c/Q_m	1.12
ICP Q_c (kN)	520
Q_c/Q_m	1.18
Fugro Q_c (kN)	710
Q_c/Q_m	1.61
NGI Q_c (kN)	509
Q_c/Q_m	1.16

Site ID No. 31: Akasaka, Tokyo, Japan.

Ref.: BCP-committee (1971): Field tests on piles in sand. Soils and Foundations, 11(2), 29–49.

Cone penetrometer data

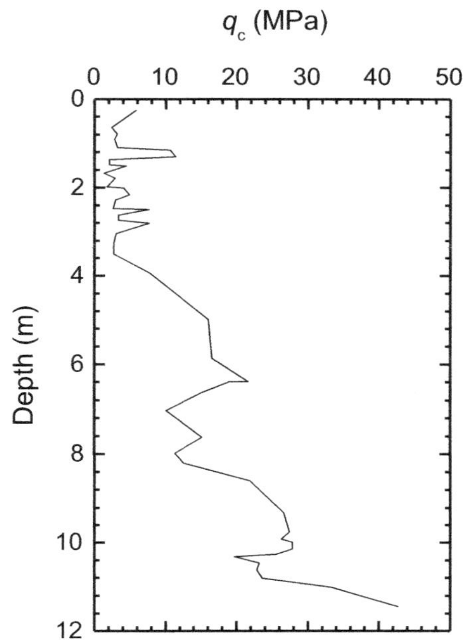

q_c (MPa)
0 10 20 30 40 50
Depth (m)
0 2 4 6 8 10 12

Detail	Description
Site name and location	Akasaka,Tokyo
Soil type (s)	Fill of loose sand, deposit of diluvial sand
Water table depth (m)	9
Pile type (s)	Closed-ended steel pipe pile
Type of cone penetrometer testing	CPT
Number of pile load tests	1
Comments	Piles installed by various means. Only the test on the single hammer-driven pile is accepted. Interface friction angle estimated with PSD, and soil unit weight applies the default value

Site ID No. 31

Pile ID: 6C

Load–displacement data

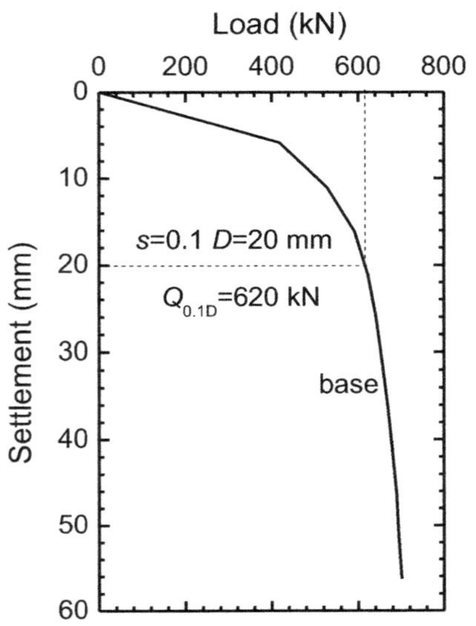

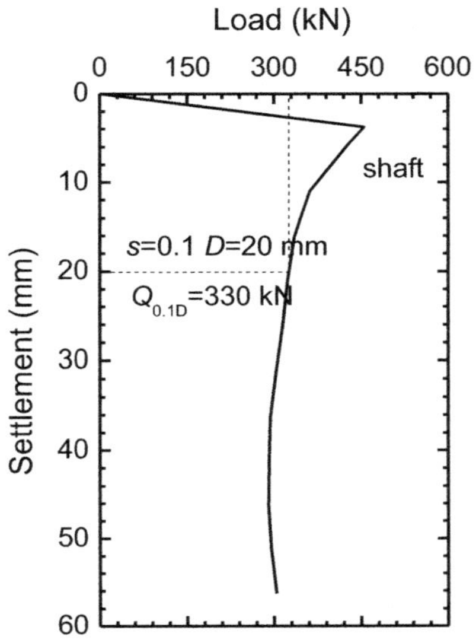

Detail	Description
Pile type/material	Closed-ended steel pipe pile
Length, L (m)	11
Outer diameter, D (mm)	200
Installation method	Driven
Set up time, days	—
Loading mode	Compression
$Q_{max\text{-}measured}$ (kN)	1500
Q_m (kN)	950
Q_s (kN)	330
Q_b (kN)	620
API Q_c (kN)	668
Q_c/Q_m	0.70
UWA Q_c (kN)	1165
Q_c/Q_m	1.23
ICP Q_c (kN)	1178
Q_c/Q_m	1.24
Fugro Q_c (kN)	970
Q_c/Q_m	1.02
NGI Q_c (kN)	1242
Q_c/Q_m	1.31

Site ID No. 32: Hound Point, Scotland.

Ref.: Williams et al. (1997): Unexpected behaviour of large diameter tubular steel piles. Proc. Int. Conf. on Foundation Failures. IES, NTU, NUS and Inst. Structural Engineers, Singapore, 363–378.

Cone penetrometer data

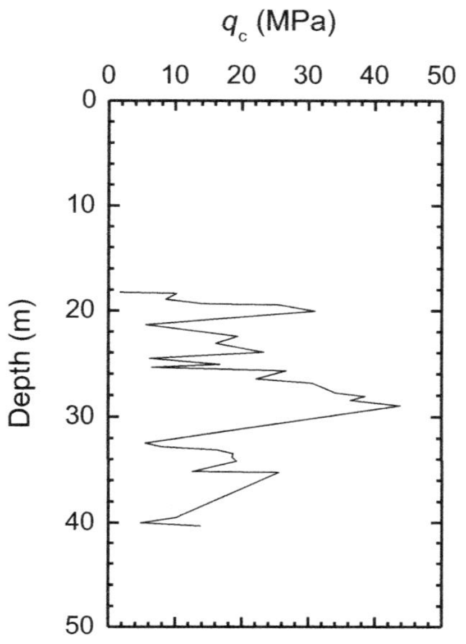

Detail	Description
Site name and location	Hound Point
Soil type (s)	Granular deposit, dense sandy gravel and cobbies unlyied very soft silty clay
Water table depth (m)	0
Pile type (s)	Three open-ended steel piles
Type of cone penetrometer testing	CPT
Number of pile load tests	3
Comments	A thick clay layer up to 17 m; one pile tested in compression, second pile tested in tension after driving to two penetrations. Piles were driven in coarse sandy gravel layer. Interface friction angle and soil unit weight apply default value.

Pile ID: P(O)-C

Load–displacement data

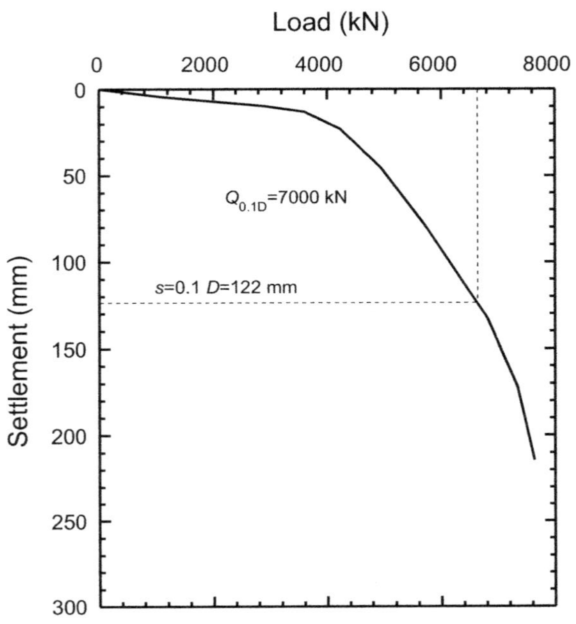

Detail	Description
Pile type/material	Open-ended steel pipe pile
Length, L (m)	26
Outer diameter, D (mm)	1220
Wall thickness, t (mm)	24.2
Installation method	Driven
Set up time, days	21
Loading mode	Compression
$Q_{max\text{-}measured}$ (kN)	7500
Q_m (kN)	7000
Q_s (kN)	4000
Q_b (kN)	3000
API Q_c (kN)	13474
Q_c/Q_m	1.92
UWA Q_c (kN)	6479
Q_c/Q_m	0.93
ICP Q_c (kN)	5126
Q_c/Q_m	0.73
Fugro Q_c (kN)	10995
Q_c/Q_m	1.57
NGI Q_c (kN)	10025
Q_c/Q_m	1.43

Pile ID: P(O)-T1

Load–displacement data	Detail	Description
Values mentioned in reference paper with no curves given	Pile type/material	Open-ended steel pipe pile
	Length, L (m)	34
	Outer diameter, D (mm)	1220
	Wall thickness, t (mm)	24.2
	Installation method	Driven
	Set up time, days	4
	Loading mode	Tension
	$Q_{max\text{-}measured}$ (kN)	3860
	Q_m (kN)	3860
	Q_s (kN)	3860
	Q_b (kN)	—
	API Q_c (kN)	4860
	Q_c/Q_m	1.26
	UWA Q_c (kN)	4201
	Q_c/Q_m	1.09
	ICP Q_c (kN)	4802
	Q_c/Q_m	1.24
	Fugro Q_c (kN)	4424
	Q_c/Q_m	1.15
	NGI Q_c (kN)	5467
	Q_c/Q_m	1.42

Pile ID: P(O)-T2

Load–displacement data	Detail	Description
Values mentioned in reference paper with no curves given	Pile type/material	Open-ended steel pipe pile
	Length, L (m)	41
	Outer diameter, D (mm)	1220
	Wall thickness, t (mm)	24.2
	Installation method	Driven
	Set up time, days	4
	Loading mode	Tension
	$Q_{max\text{-}measured}$ (kN)	3740
	Q_m (kN)	3740
	Q_s (kN)	3740
	Q_b (kN)	—
	API Q_c (kN)	7468
	Q_c/Q_m	2.00
	UWA Q_c (kN)	4590
	Q_c/Q_m	1.23
	ICP Q_c (kN)	5500
	Q_c/Q_m	1.47
	Fugro Q_c (kN)	4012
	Q_c/Q_m	1.07
	NGI Q_c (kN)	6377
	Q_c/Q_m	1.71

Site ID No. 33: Leman BD, North Sea.

Ref.: Jardine et al. (1998): Axial capacity of offshore piles in dense North Sea Sand. Journal of Geotechnical and Geoenvironmental Engineering, ASCE, 124(2), 171–178.

Cone penetrometer data

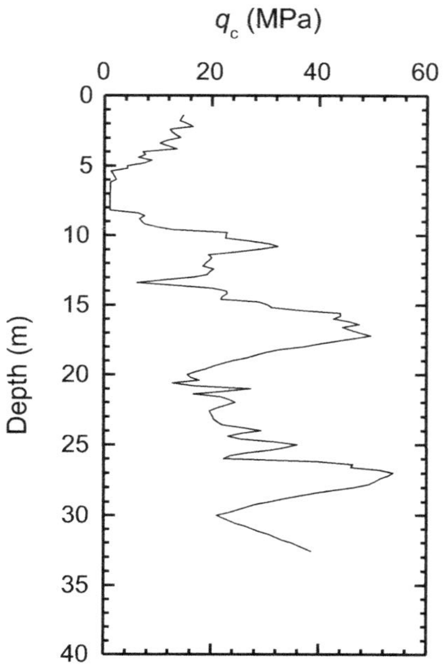

Detail	Description
Site name and location	Leman field, southern North Sea
Soil type (s)	Medium dense silty fine sand, dense fine sand
Water table depth (m)	0
Pile type (s)	Open-ended steel pipe pile
Type of cone penetrometer testing	CPT
Number of pile load tests	1
Comments	A gas conductor tested in tension from a platform in the Leman field, southern North Sea. Interface friction angle and soil unit weight apply default value.

Pile ID: BD

Load–displacement data

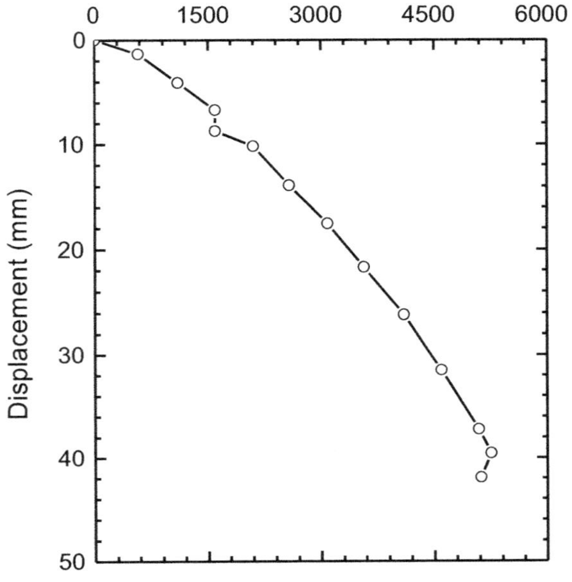

Load (kN)
0 1500 3000 4500 6000
Displacement (mm)
0 10 20 30 40 50

Detail	Description
Pile type/material	Open-ended steel pipe pile
Length, L (m)	38.1
Outer diameter, D (mm)	660
Wall thickness, t (mm)	19
Installation method	Driven
Set up time, days	—
Loading mode	Tension
$Q_{max\text{-}measured}$ (kN)	5250
Q_m (kN)	5250
Q_s (kN)	5250
Q_b (kN)	—
API Q_c (kN)	5617
Q_c/Q_m	1.07
UWA Q_c (kN)	5350
Q_c/Q_m	1.02
ICP Q_c (kN)	5382
Q_c/Q_m	1.03
Fugro Q_c (kN)	3736
Q_c/Q_m	0.71
NGI Q_c (kN)	6604
Q_c/Q_m	1.26

Site ID No. 34: Baghdad University, Iraq.

Ref.: Altaee et al. (1992): Axial load transfer for piles in sand. I. Tests on an instrumented precast pile. Canadian Geotechnical Journal, 29(1), 11–20.

Cone penetrometer data

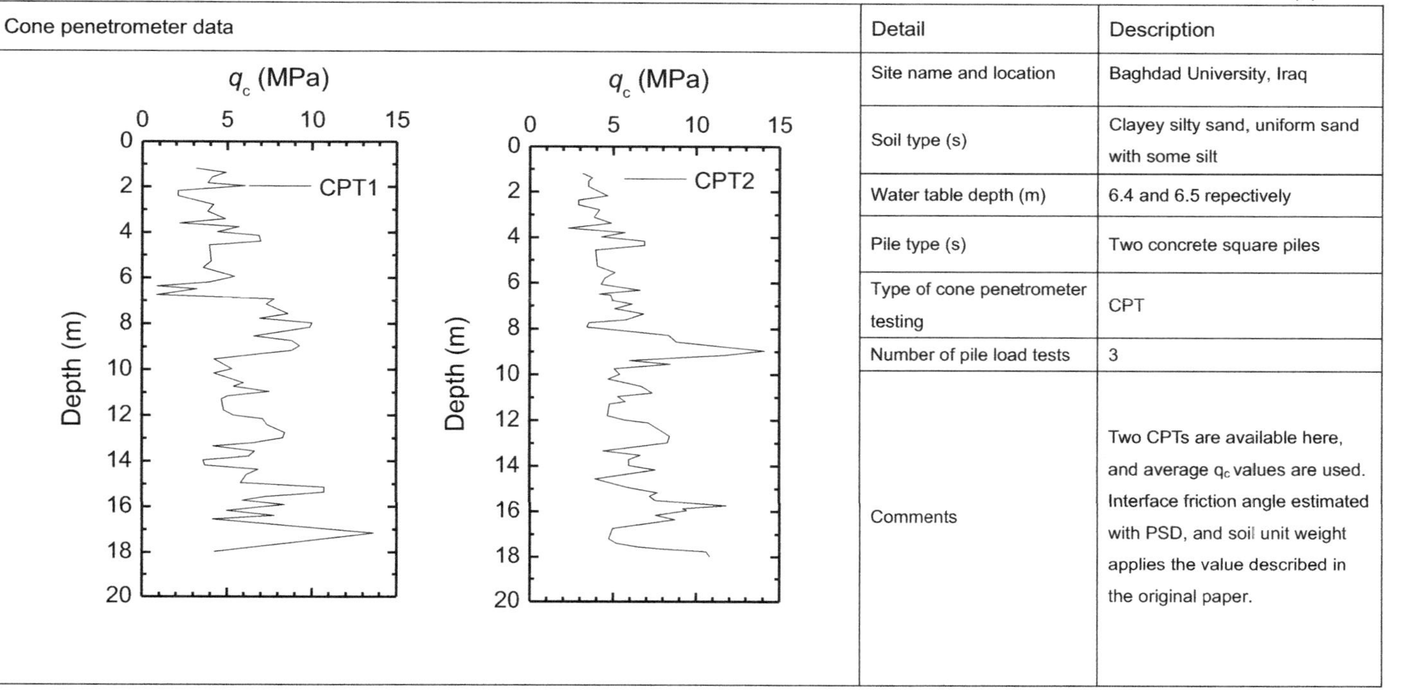

Detail	Description
Site name and location	Baghdad University, Iraq
Soil type (s)	Clayey silty sand, uniform sand with some silt
Water table depth (m)	6.4 and 6.5 repectively
Pile type (s)	Two concrete square piles
Type of cone penetrometer testing	CPT
Number of pile load tests	3
Comments	Two CPTs are available here, and average q_c values are used. Interface friction angle estimated with PSD, and soil unit weight applies the value described in the original paper.

Site ID No. 34

Pile ID: P1-C

Load–displacement data

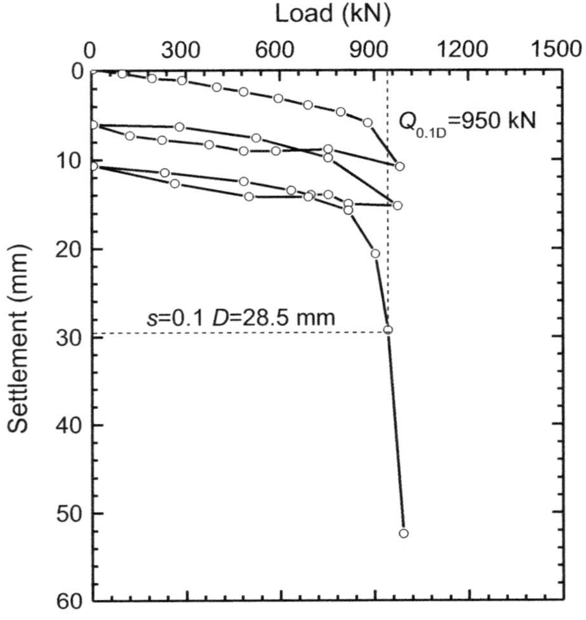

Detail	Description
Pile type/material	Square concrete pile
Length, L (m)	11
Outer width, B (mm)	253
Installation method	Driven
Set up time, days	88
Loading mode	Compression
$Q_{max\text{-}measured}$ (kN)	1100
Q_m (kN)	950
Q_s (kN)	580
Q_b (kN)	370
API Q_c (kN)	479
Q_c/Q_m	0.50
UWA Q_c (kN)	596
Q_c/Q_m	0.63
ICP Q_c (kN)	519
Q_c/Q_m	0.55
Fugro Q_c (kN)	690
Q_c/Q_m	0.73
NGI Q_c (kN)	598
Q_c/Q_m	0.63

Pile ID: P1-T

Load–displacement data

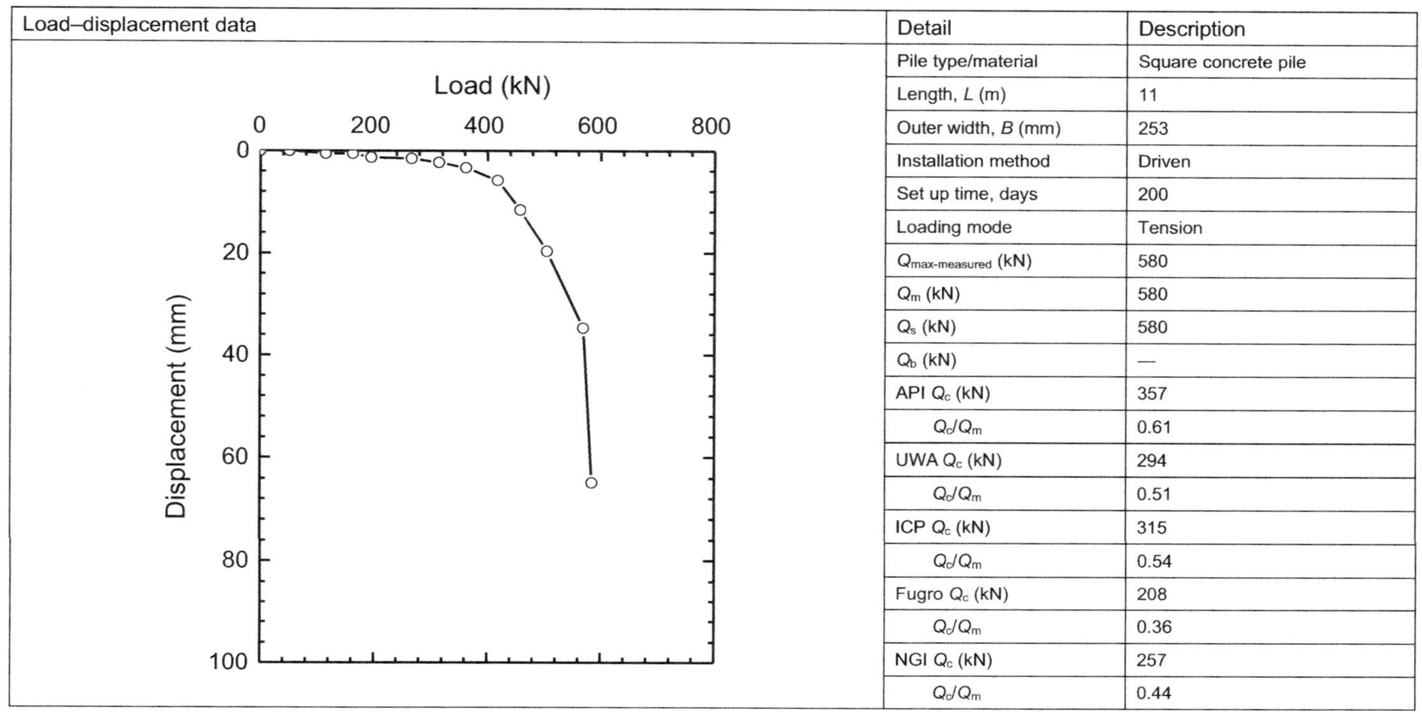

Detail	Description
Pile type/material	Square concrete pile
Length, L (m)	11
Outer width, B (mm)	253
Installation method	Driven
Set up time, days	200
Loading mode	Tension
$Q_{max\text{-}measured}$ (kN)	580
Q_m (kN)	580
Q_s (kN)	580
Q_b (kN)	—
API Q_c (kN)	357
Q_c/Q_m	0.61
UWA Q_c (kN)	294
Q_c/Q_m	0.51
ICP Q_c (kN)	315
Q_c/Q_m	0.54
Fugro Q_c (kN)	208
Q_c/Q_m	0.36
NGI Q_c (kN)	257
Q_c/Q_m	0.44

Site ID No. 34

Pile ID: P2-C

Load–displacement data

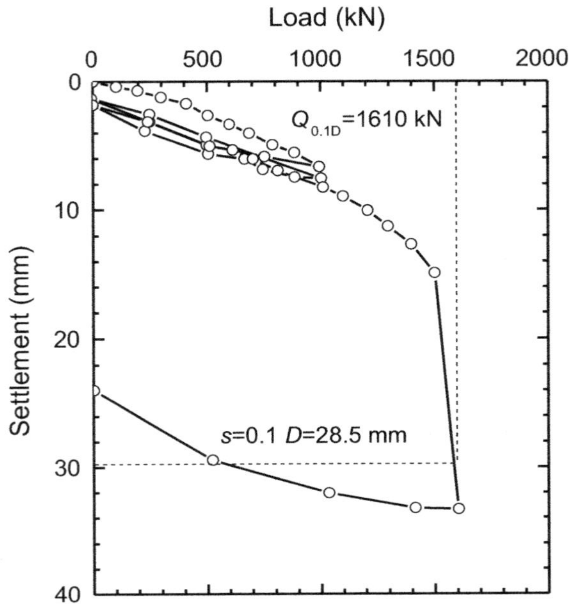

Detail	Description
Pile type/material	Square concrete pile
Length, L (m)	15
Outer width, B (mm)	253
Installation method	Driven
Set up time, days	42
Loading mode	Compression
$Q_{max\text{-}measured}$ (kN)	1610
Q_m (kN)	1610
Q_s (kN)	1251
Q_b (kN)	359
API Q_c (kN)	710
Q_c/Q_m	0.44
UWA Q_c (kN)	756
Q_c/Q_m	0.47
ICP Q_c (kN)	754
Q_c/Q_m	0.47
Fugro Q_c (kN)	720
Q_c/Q_m	0.45
NGI Q_c (kN)	661
Q_c/Q_m	0.41

Site ID No. 35: Dunkirk CLAROM, France.

Ref.: Chow (1997): Investigations into displacement pile behaviour for offshore foundations. (Ph.D. thesis), Imperial College London, London.

Cone penetrometer data

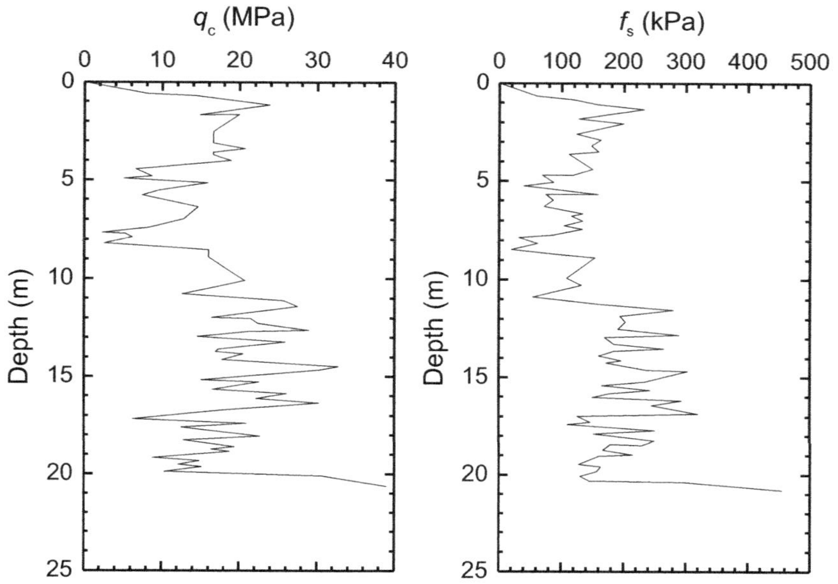

Detail	Description
Site name and location	Zip Les Hunttes site,west of Dunkirk
Soil type (s)	Hydraulic fill and Flandrian sand
Water table depth (m)	4
Pile type (s)	Open-ended steel pipe piles
Type of cone penetrometer testing	CPT
Number of pile load tests	2
Comments	Two piles accepted are from CLAROM project. Pile CL was subjected to tension loading first, then to compression. Interface friction angle measured with ring shear test, and unit weight applies the value in the original paper.

Site ID No. 35

Pile ID: CL-T

Load–displacement data

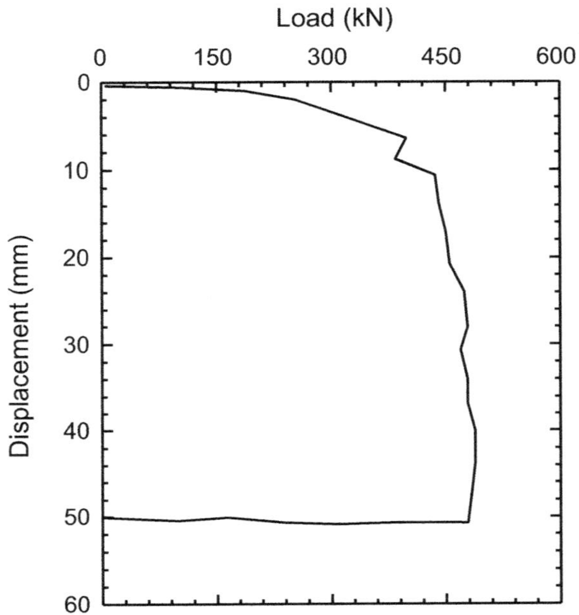

Detail	Description
Pile type/material	Open-ended steel pipe pile
Length, L (m)	11.3
Outer diameter, D/B (mm)	324
Wall thickness, t (mm)	12.7
Installation method	Driven
Set up time, days	175
Loading mode	Tension
$Q_{max\text{-}measured}$ (kN)	444
Q_m (kN)	444
Q_s (kN)	444
Q_b (kN)	—
API Q_c (kN)	451
Q_c/Q_m	1.02
UWA Q_c (kN)	527
Q_c/Q_m	1.19
ICP Q_c (kN)	481
Q_c/Q_m	1.08
Fugro Q_c (kN)	690
Q_c/Q_m	1.55
NGI Q_c (kN)	582
Q_c/Q_m	1.31

Pile ID: CS-T

Load–displacement data

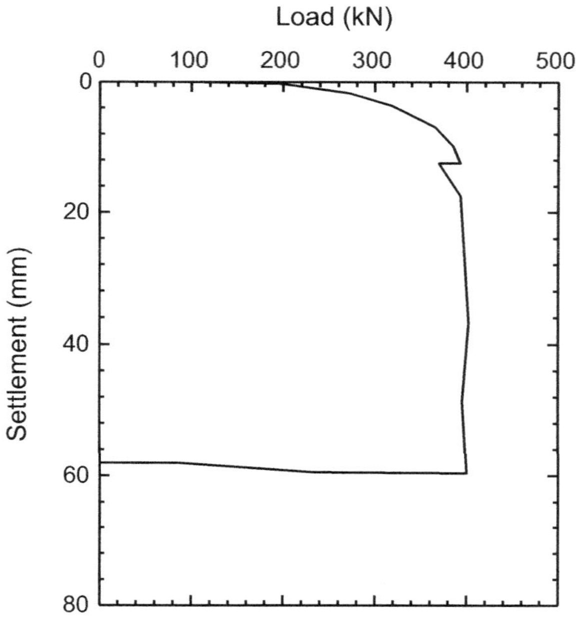

Detail	Description
Pile type/material	Open-ended steel pipe pile
Length, L (m)	11.3
Outer diameter, D (mm)	324
Wall thickness, t (mm)	12.7
Installation method	Driven
Set up time, days	187
Loading mode	Compression
$Q_{max\text{-}measured}$ (kN)	400
Q_m (kN)	400
Q_s (kN)	400
Q_b (kN)	—
API Q_c (kN)	451
Q_c/Q_m	1.13
UWA Q_c (kN)	527
Q_c/Q_m	1.32
ICP Q_c (kN)	481
Q_c/Q_m	1.20
Fugro Q_c (kN)	690
Q_c/Q_m	1.73
NGI Q_c (kN)	582
Q_c/Q_m	1.46

Site ID No. 36: Dunkirk GOPAL, France.

Ref.: Jardine et al. (2006): Some observations of the effects of time on the capacity of piles driven in sand. Géotechnique, 56(4), 227–244.

Cone penetrometer data

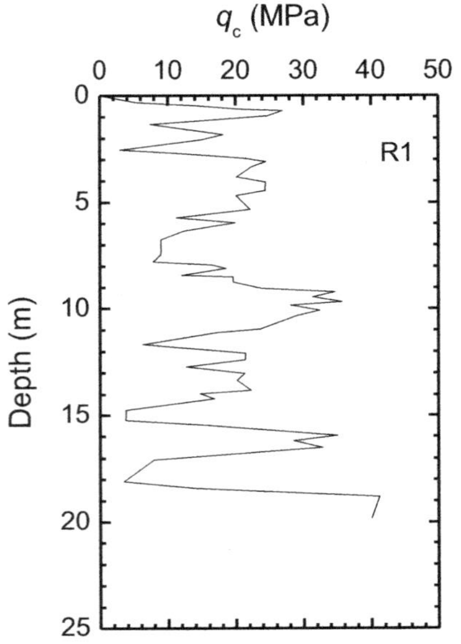

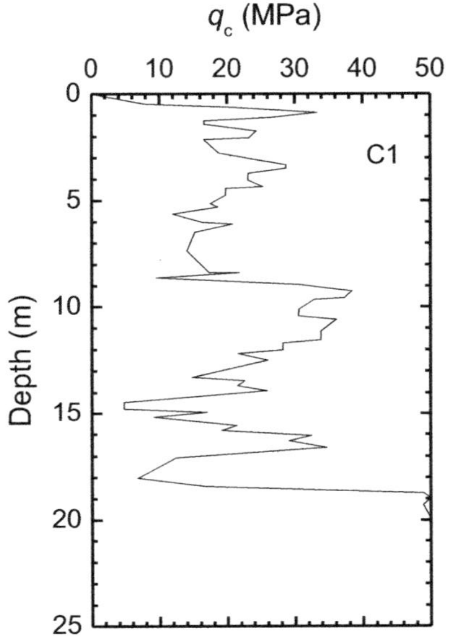

Detail	Description
Site name and location	Zip Les Hunttes site,west of Dunkirk
Soil type (s)	Hydraulic fill, Flandrian sand
Water table depth (m)	4
Pile type (s)	Two open-ended steel pipe piles
Type of cone penetrometer testing	CPT
Number of pile load tests	3
Comments	Altogether seven piles in GOPAL project. Only three are accepted, and all the other piles were subjected to cycling loading first before rapid tension. Interpace friction angle measured with ring shear test, and unit weight applies the value in the original paper.

Pile ID: R1-T

Load–displacement data

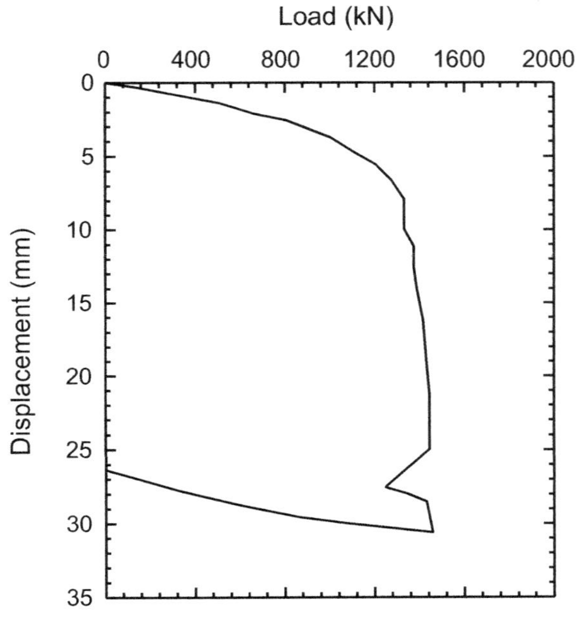

Detail	Description
Pile type/material	Open-ended steel pipe pile
Length, L (m)	19.3
Outer diameter, D (mm)	457
Wall thickness, t (mm)	13.5
Installation method	Driven
Set up time, days	9
Loading mode	Tension
$Q_{max\text{-}measured}$ (kN)	1450
Q_m (kN)	1450
Q_s (kN)	1450
Q_b (kN)	—
API Q_c (kN)	1426
Q_c/Q_m	0.98
UWA Q_c (kN)	1304
Q_c/Q_m	0.90
ICP Q_c (kN)	1310
Q_c/Q_m	0.90
Fugro Q_c (kN)	1101
Q_c/Q_m	0.76
NGI Q_c (kN)	1559
Q_c/Q_m	1.08

Pile ID: C1-C

Load–displacement data

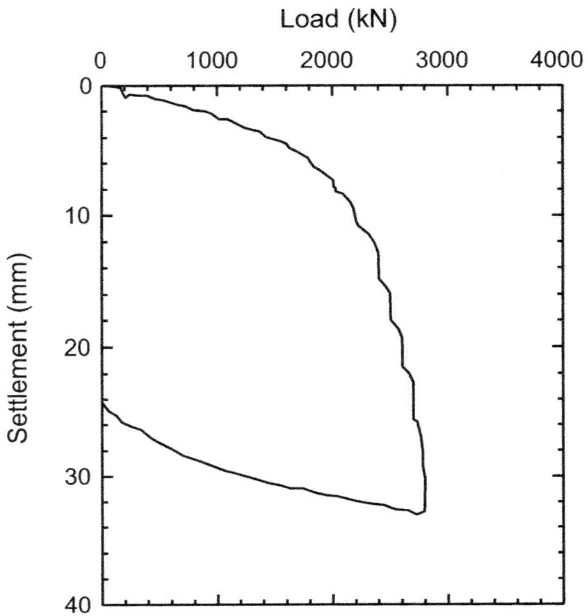

Load (kN)
0 1000 2000 3000 4000
Settlement (mm)
0 10 20 30 40

Detail	Description
Pile type/material	Open-ended steel pipe pile
Length, L (m)	10
Outer diameter, D (mm)	457
Wall thickness, t (mm)	13.5
Installation method	Driven
Set up time, days	68
Loading mode	Compression
$Q_{max\text{-}measured}$ (kN)	2820
Q_m (kN)	2820
Q_s (kN)	Not isolated
Q_b (kN)	Not isolated
API Q_c (kN)	1457
Q_c/Q_m	0.52
UWA Q_c (kN)	2657
Q_c/Q_m	0.94
ICP Q_c (kN)	2614
Q_c/Q_m	0.93
Fugro Q_c (kN)	2777
Q_c/Q_m	0.98
NGI Q_c (kN)	2361
Q_c/Q_m	0.84

Pile ID: C1-T

Load–displacement data

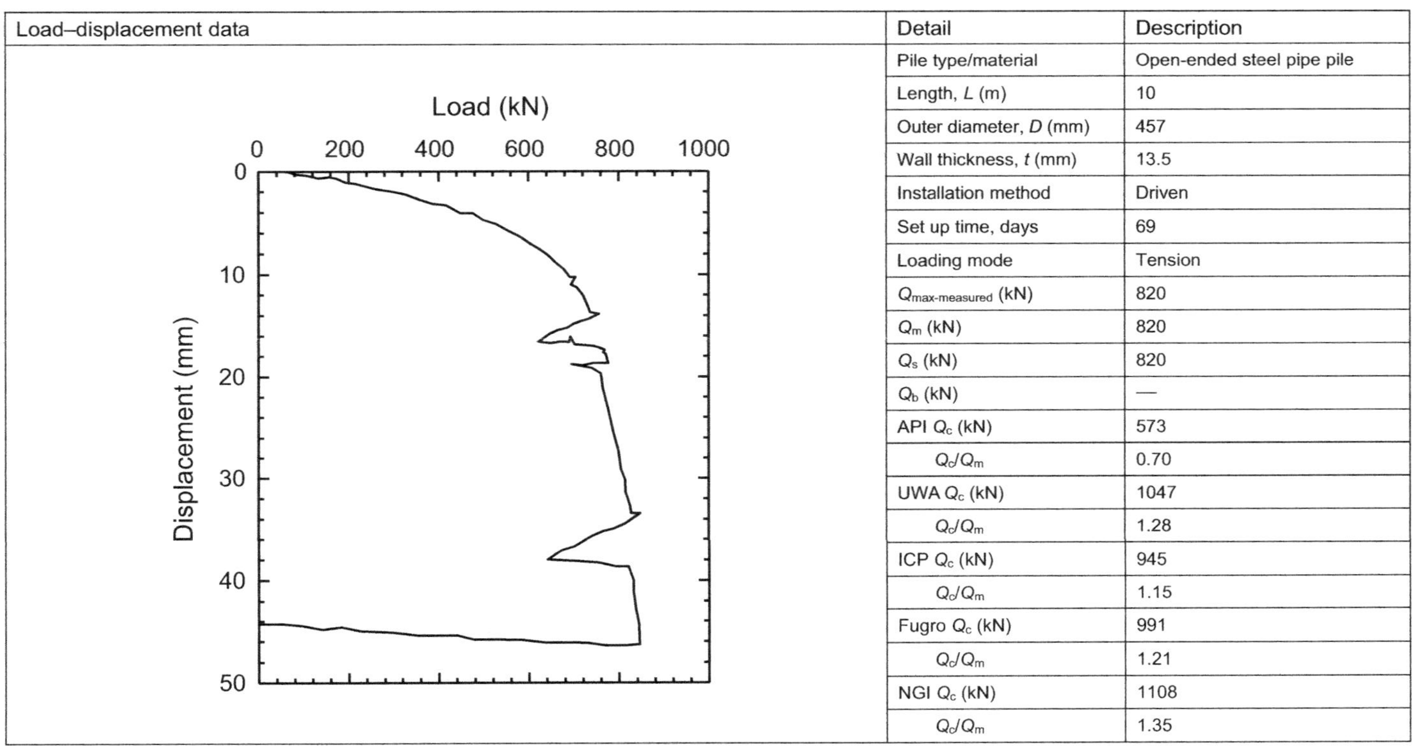

Detail	Description
Pile type/material	Open-ended steel pipe pile
Length, L (m)	10
Outer diameter, D (mm)	457
Wall thickness, t (mm)	13.5
Installation method	Driven
Set up time, days	69
Loading mode	Tension
$Q_{max\text{-}measured}$ (kN)	820
Q_m (kN)	820
Q_s (kN)	820
Q_b (kN)	—
API Q_c (kN)	573
Q_c/Q_m	0.70
UWA Q_c (kN)	1047
Q_c/Q_m	1.28
ICP Q_c (kN)	945
Q_c/Q_m	1.15
Fugro Q_c (kN)	991
Q_c/Q_m	1.21
NGI Q_c (kN)	1108
Q_c/Q_m	1.35

Site ID No. 37: Euripides, The Netherlands

Ref.: Kolk et al. (2005b): Results of axial load tests on pipe piles in very dense sands: The EURIPIDES JIP Proc. Int. Symp. on Frontiers in Offshore Geomechanics, ISFOG, Taylor & Francis, London, 661–667.

Cone penetrometer data

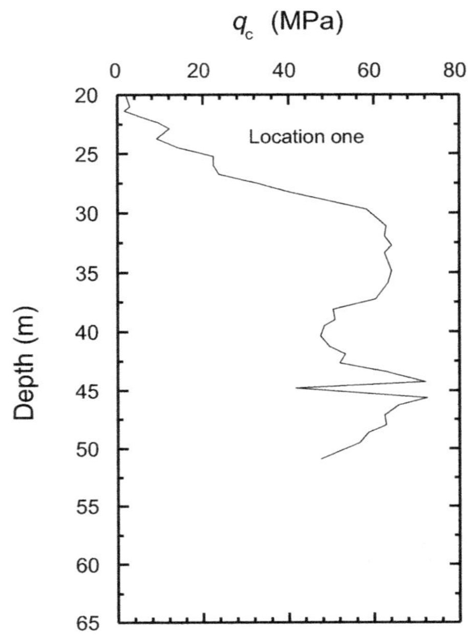

Detail	Description
Site name and location	Euripides Joint Industry Project, The Netherlands
Soil type (s)	Very dense sand, silty and clay, Low strength Holocene material overlying dense sands.
Water table depth (m)	1
Pile type (s)	One pile which was installed once, then extracted and re-driven
Type of cone penetrometer testing	CPT
Number of pile load tests	6
Comments	Same instrumented pile driven at two locations and tested after penetrations to four depths at this location. Loads measured from 22 m level and settlements corrected for compression in steel between this level and ground surface.

Pile ID: Ia

Load–displacement data

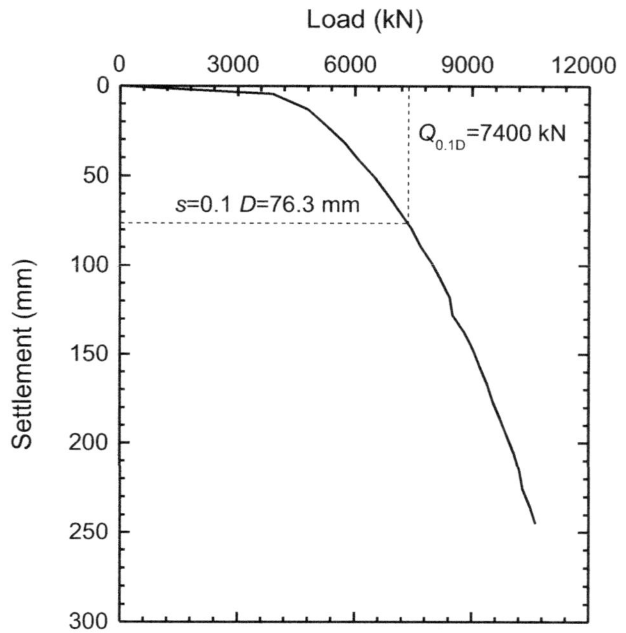

Detail	Description
Pile type/material	Open-ended steel pipe pile
Length, L (m)	30.5
Outer diameter, D (mm)	763
Wall thickness, t (mm)	35.6
Installation method	Driven
Set up time, days	7
Loading mode	Compression
$Q_{max\text{-}measured}$ (kN)	11600
Q_m (kN)	7400
Q_s (kN)	3400
Q_b (kN)	4000
API Q_c (kN)	7385
Q_c/Q_m	1.00
UWA Q_c (kN)	10149
Q_c/Q_m	1.37
ICP Q_c (kN)	8745
Q_c/Q_m	1.18
Fugro Q_c (kN)	11091
Q_c/Q_m	1.50
NGI Q_c (kN)	7515
Q_c/Q_m	1.02

Pile ID: Ib

Load–displacement data

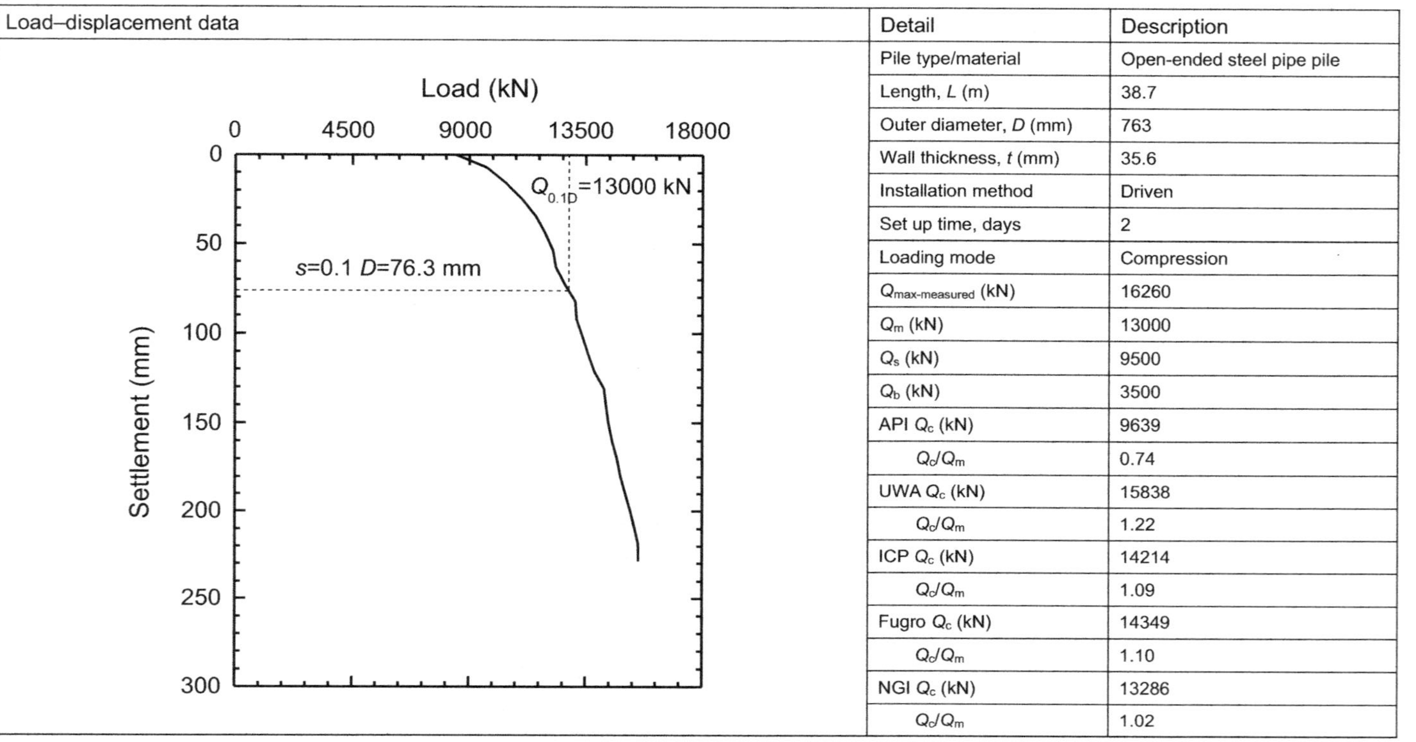

Detail	Description
Pile type/material	Open-ended steel pipe pile
Length, L (m)	38.7
Outer diameter, D (mm)	763
Wall thickness, t (mm)	35.6
Installation method	Driven
Set up time, days	2
Loading mode	Compression
$Q_{max\text{-}measured}$ (kN)	16260
Q_m (kN)	13000
Q_s (kN)	9500
Q_b (kN)	3500
API Q_c (kN)	9639
Q_c/Q_m	0.74
UWA Q_c (kN)	15838
Q_c/Q_m	1.22
ICP Q_c (kN)	14214
Q_c/Q_m	1.09
Fugro Q_c (kN)	14349
Q_c/Q_m	1.10
NGI Q_c (kN)	13286
Q_c/Q_m	1.02

Pile ID: Ic

Load–displacement data

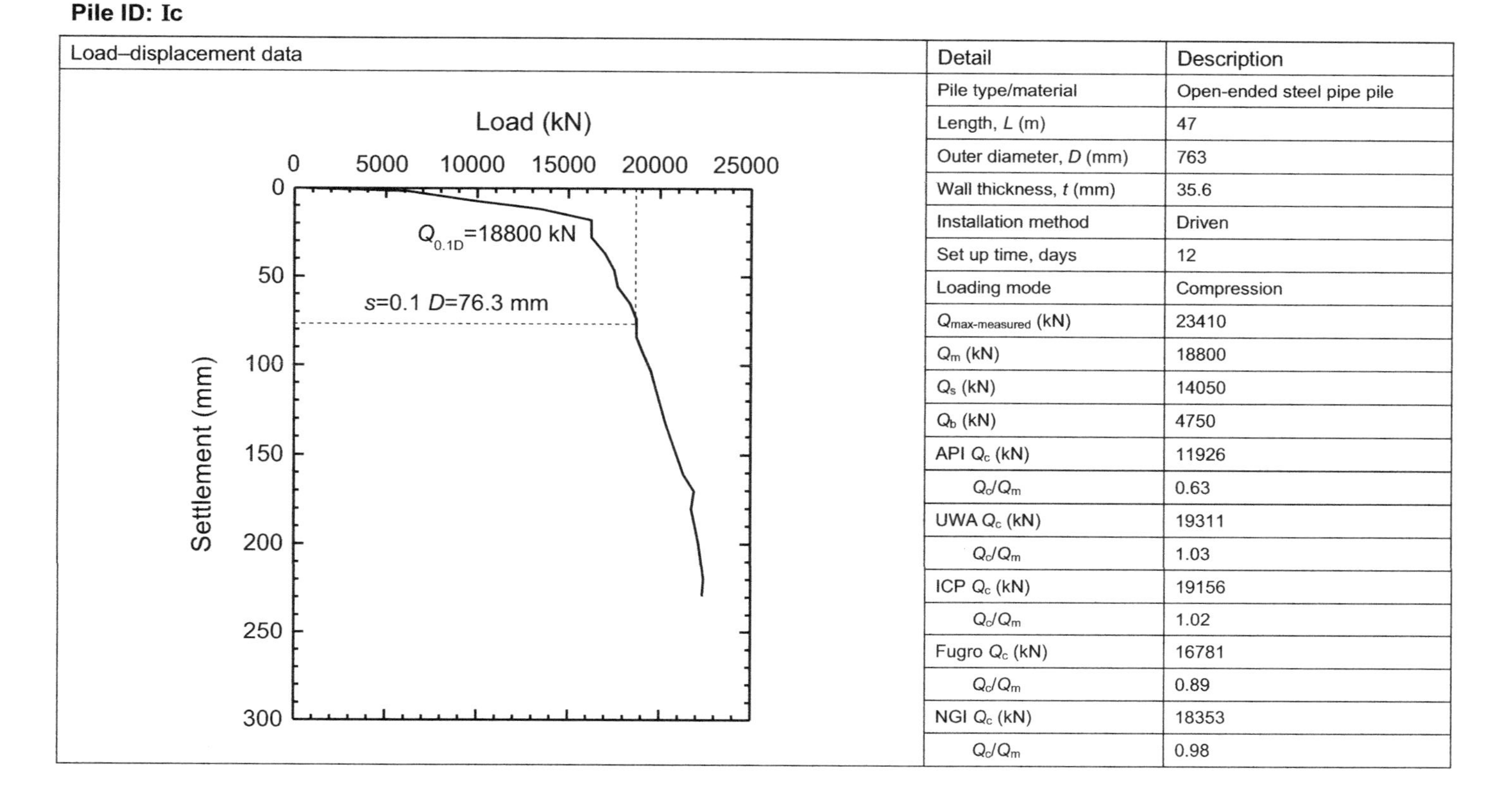

Detail	Description
Pile type/material	Open-ended steel pipe pile
Length, L (m)	47
Outer diameter, D (mm)	763
Wall thickness, t (mm)	35.6
Installation method	Driven
Set up time, days	12
Loading mode	Compression
$Q_{max\text{-}measured}$ (kN)	23410
Q_m (kN)	18800
Q_s (kN)	14050
Q_b (kN)	4750
API Q_c (kN)	11926
Q_c/Q_m	0.63
UWA Q_c (kN)	19311
Q_c/Q_m	1.03
ICP Q_c (kN)	19156
Q_c/Q_m	1.02
Fugro Q_c (kN)	16781
Q_c/Q_m	0.89
NGI Q_c (kN)	18353
Q_c/Q_m	0.98

Site ID No. 37

Pile ID: Ia-T

Load–displacement data

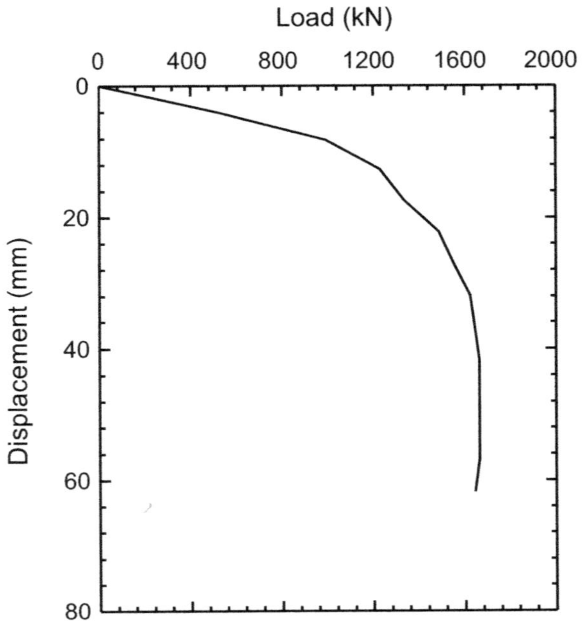

Detail	Description
Pile type/material	Open-ended steel pipe pile
Length, L (m)	30.5
Outer diameter, D (mm)	763
Wall thickness, t (mm)	35.6
Installation method	Driven
Set up time, days	7
Loading mode	Tension
$Q_{max\text{-}measured}$ (kN)	1660
Q_m (kN)	1660
Q_s (kN)	1660
Q_b (kN)	—
API Q_c (kN)	1814
Q_c/Q_m	1.09
UWA Q_c (kN)	2751
Q_c/Q_m	1.66
ICP Q_c (kN)	2804
Q_c/Q_m	1.69
Fugro Q_c (kN)	4295
Q_c/Q_m	2.59
NGI Q_c (kN)	2768
Q_c/Q_m	1.67

Pile ID: Ib-T

Load–displacement data

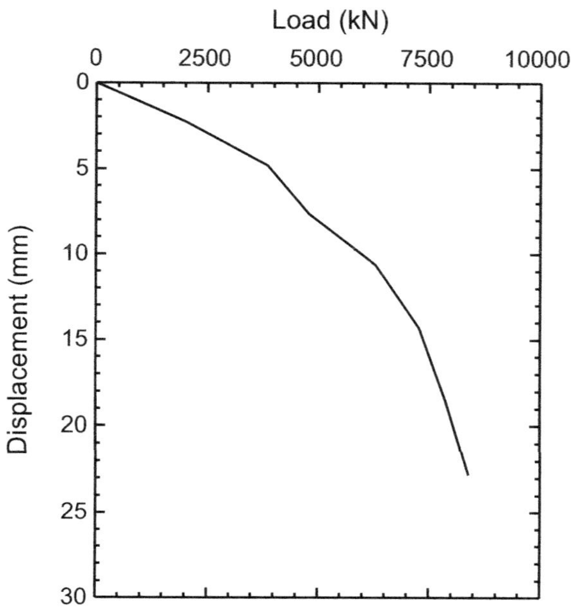

Detail	Description
Pile type/material	Open-ended steel pipe pile
Length, L (m)	38.7
Outer diameter, D (mm)	763
Wall thickness, t (mm)	35.6
Installation method	Driven
Set up time, days	2
Loading mode	Tension
$Q_{max\text{-}measured}$ (kN)	8400
Q_m (kN)	8400
Q_s (kN)	8400
Q_b (kN)	—
API Q_c (kN)	4103
Q_c/Q_m	0.49
UWA Q_c (kN)	6932
Q_c/Q_m	0.83
ICP Q_c (kN)	7317
Q_c/Q_m	0.87
Fugro Q_c (kN)	7052
Q_c/Q_m	0.84
NGI Q_c (kN)	7209
Q_c/Q_m	0.86

Pile ID: Ic-T

Load–displacement data

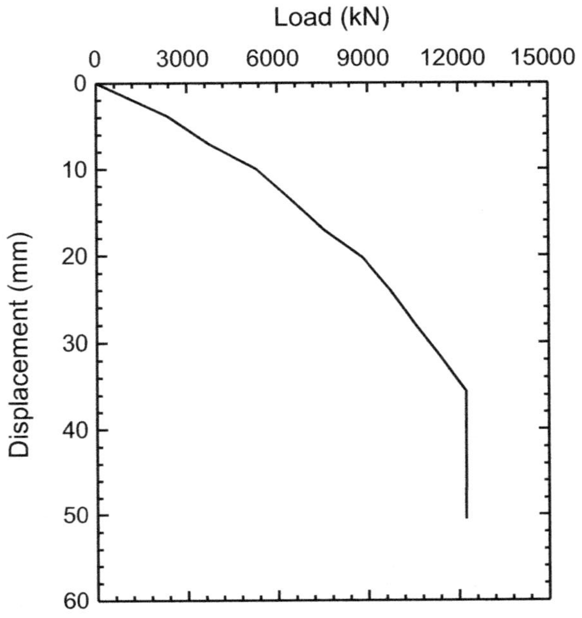

Detail	Description
Pile type/material	Open-ended steel pipe pile
Length, L (m)	47
Outer diameter, D (mm)	763
Wall thickness, t (mm)	35.6
Installation method	Driven
Set up time, days	12
Loading mode	Tension
$Q_{max\text{-}measured}$ (kN)	12500
Q_m (kN)	12500
Q_s (kN)	12500
Q_b (kN)	—
API Q_c (kN)	6440
Q_c/Q_m	0.52
UWA Q_c (kN)	9046
Q_c/Q_m	0.72
ICP Q_c (kN)	10107
Q_c/Q_m	0.81
Fugro Q_c (kN)	8894
Q_c/Q_m	0.71
NGI Q_c (kN)	10552
Q_c/Q_m	0.84

Site ID No. 38: Euripides, The Netherlands.

Ref.: Kolk et al. (2005b): Results of axial load tests on pipe piles in very dense sands: The EURIPIDES JIP Proc. Int. Symp. on Frontiers in Offshore Geomechanics, ISFOG, Taylor & Francis, London, 661–667.

Cone penetrometer data

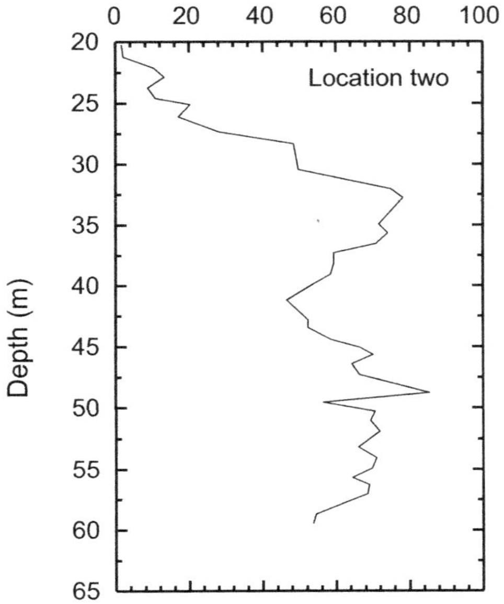

Detail	Description
Site name and location	Euripides Joint Industry Project, The Netherlands
Soil type (s)	Very dense sand, silty and clay
Water table depth (m)	1
Pile type (s)	One pile which was installed once, then extracted and re-driven.
Type of cone penetrometer testing	CPT
Number of pile load tests	2
Comments	Same instrumented pile driven at two locations. Low strength Holocene material overlying dense sands. Loads measured from 22 m level and settlements corrected for compression in steel between this level and ground surface.

Site ID No. 38

Pile ID: II

Load–displacement data

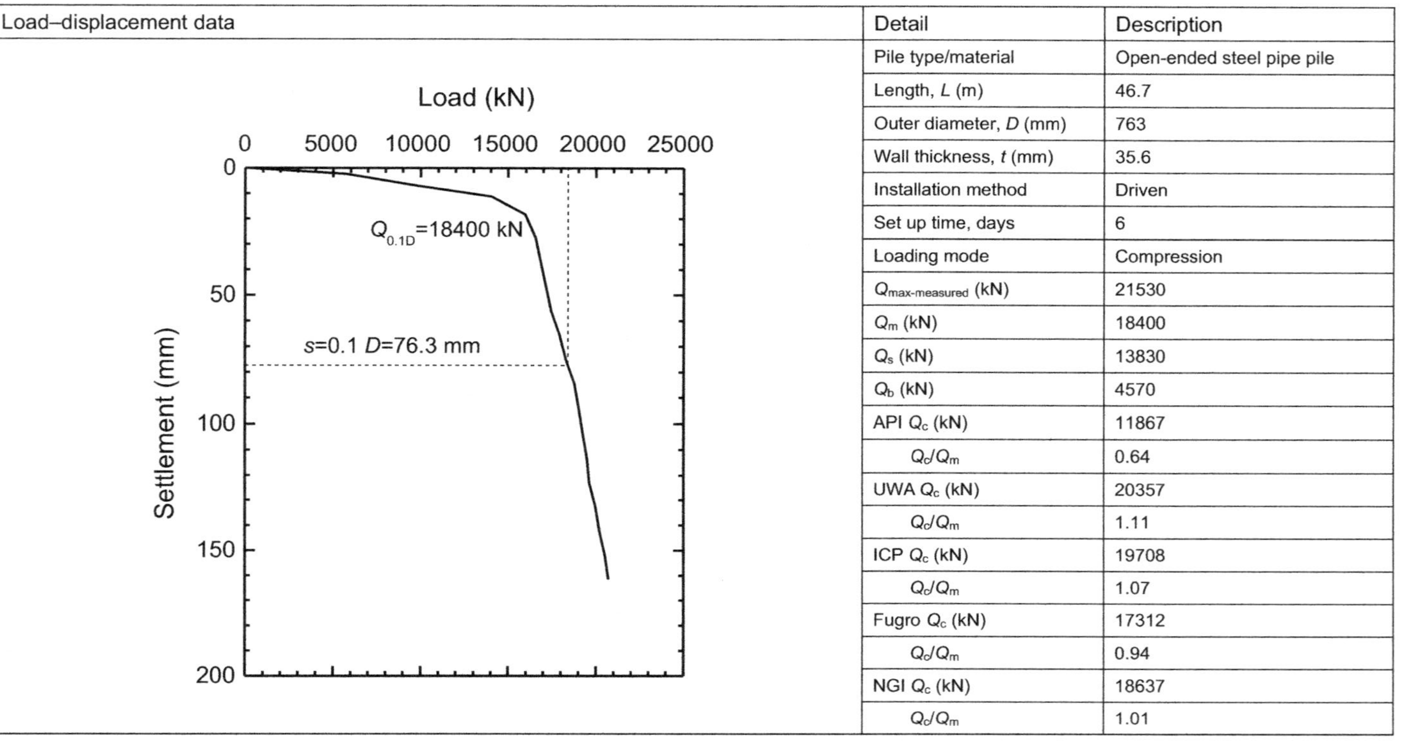

Detail	Description
Pile type/material	Open-ended steel pipe pile
Length, L (m)	46.7
Outer diameter, D (mm)	763
Wall thickness, t (mm)	35.6
Installation method	Driven
Set up time, days	6
Loading mode	Compression
$Q_{max\text{-}measured}$ (kN)	21530
Q_m (kN)	18400
Q_s (kN)	13830
Q_b (kN)	4570
API Q_c (kN)	11867
Q_c/Q_m	0.64
UWA Q_c (kN)	20357
Q_c/Q_m	1.11
ICP Q_c (kN)	19708
Q_c/Q_m	1.07
Fugro Q_c (kN)	17312
Q_c/Q_m	0.94
NGI Q_c (kN)	18637
Q_c/Q_m	1.01

Pile ID: II-T

Load–displacement data

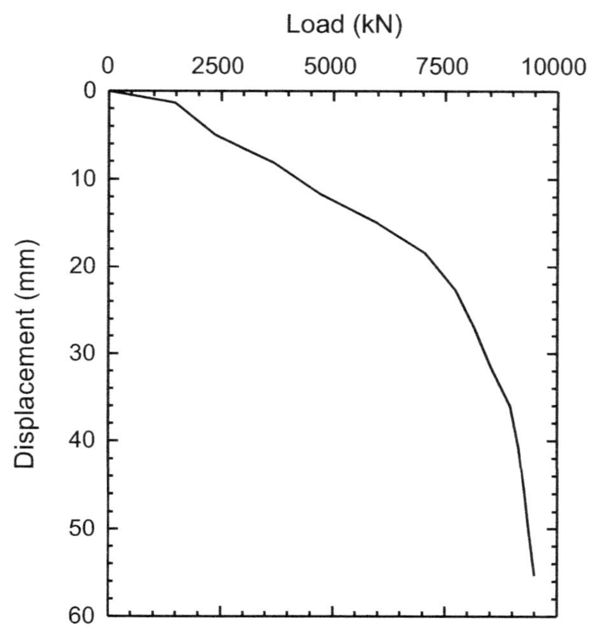

Detail	Description
Pile type/material	Open-ended steel pipe pile
Length, L (m)	46.7
Outer diameter, D (mm)	763
Wall thickness, t (mm)	35.6
Installation method	Driven
Set up time, days	6
Loading mode	Tension
$Q_{max\text{-}measured}$ (kN)	9500
Q_m (kN)	9500
Q_s (kN)	9500
Q_b (kN)	—
API Q_c (kN)	6319
Q_c/Q_m	0.67
UWA Q_c (kN)	9262
Q_c/Q_m	0.97
ICP Q_c (kN)	10377
Q_c/Q_m	1.09
Fugro Q_c (kN)	9039
Q_c/Q_m	0.95
NGI Q_c (kN)	10740
Q_c/Q_m	1.13

Site ID No. 39: Locks and Dam, USA.

Ref.: Briaud et al. (1989b): Analysis of pile load tests at Lock and Dam 26. Foundation engineering: Current principles and practices, Evanston, ACSE, Reston, VA, 925–942.

Cone penetrometer data

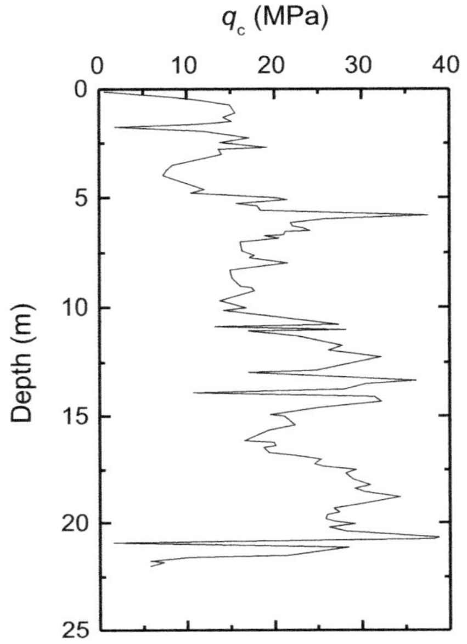

q_c (MPa)
0 10 20 30 40
Depth (m)
0 5 10 15 20 25

Detail	Description
Site name and location	The new Locks and Dam 26, Mississippi river, Alton
Soil type (s)	Alluvial deposits,recent alluvium and alluvial outwash, glacial deposits
Water table depth (m)	0
Pile type (s)	Six closed-ended steel piles
Type of cone penetrometer testing	CPTU with pore pressure
Number of pile load tests	6
Comments	Briaud et al., report 28 load tests, only six piles in Group 3 pass the quality criteria because the others are H piles. Interface friction angle estimated with PSD, and soil unit weight applies default value.

Pile ID: 3-1

Load–displacement data

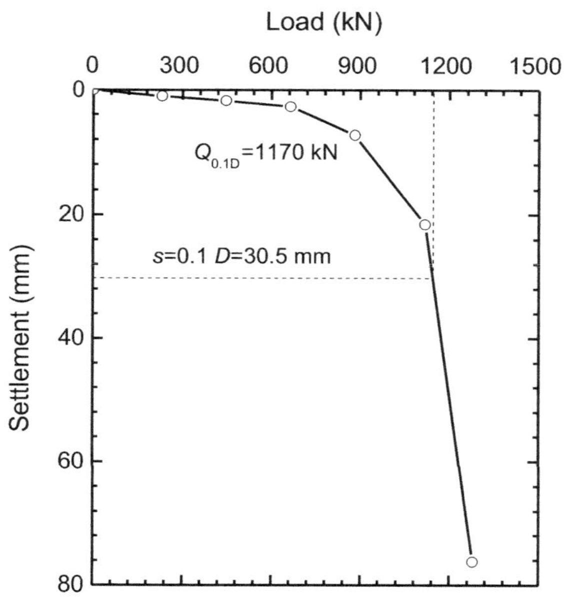

Detail	Description
Pile type/material	Closed-ended steel pipe pile
Length, L (m)	14.2
Outer diameter, D (mm)	305
Installation method	Driven
Set up time, days	35
Loading mode	Compression
$Q_{max\text{-}measured}$ (kN)	1320
Q_m (kN)	1170
Q_s (kN)	Not isolated
Q_b (kN)	Not isolated
API Q_c (kN)	958
Q_c/Q_m	0.82
UWA Q_c (kN)	2049
Q_c/Q_m	1.75
ICP Q_c (kN)	2000
Q_c/Q_m	1.71
Fugro Q_c (kN)	2240
Q_c/Q_m	1.91
NGI Q_c (kN)	3036
Q_c/Q_m	2.59

Site ID No. 39

Pile ID: 3-4

Load–displacement data

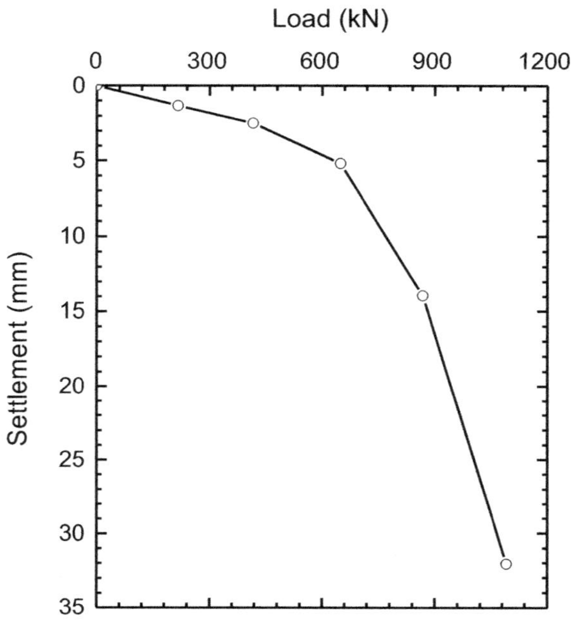

Load (kN)
0 300 600 900 1200
Settlement (mm)
0 5 10 15 20 25 30 35

Detail	Description
Pile type/material	Closed-ended steel pipe pile
Length, L (m)	14.4
Outer diameter, D (mm)	356
Installation method	Driven
Set up time, days	27
Loading mode	Compression
$Q_{max\text{-}measured}$ (kN)	1130
Q_m (kN)	1150
Q_s (kN)	Not isolated
Q_b (kN)	Not isolated
API Q_c (kN)	1192
Q_c/Q_m	1.04
UWA Q_c (kN)	2615
Q_c/Q_m	2.23
ICP Q_c (kN)	2430
Q_c/Q_m	2.11
Fugro Q_c (kN)	2950
Q_c/Q_m	2.57
NGI Q_c (kN)	3746
Q_c/Q_m	3.26

Pile ID: 3-7

Load–displacement data

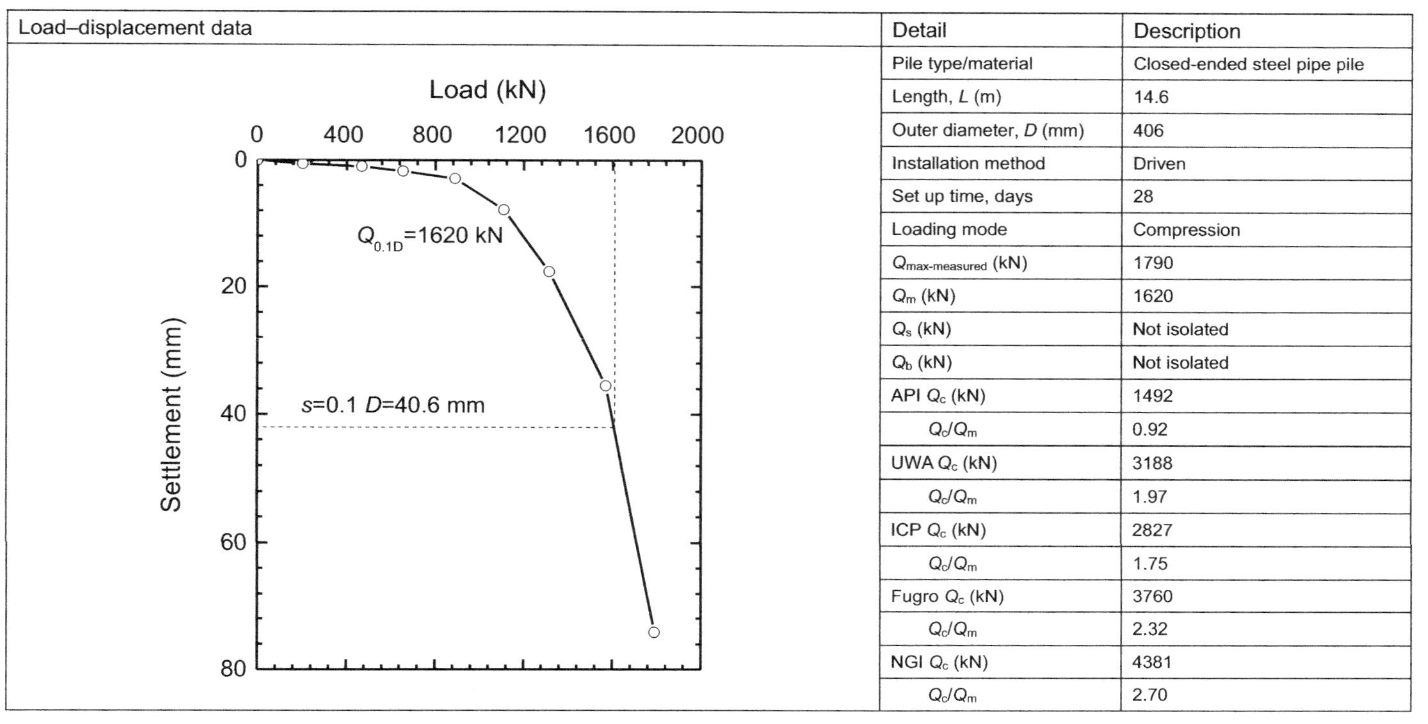

Detail	Description
Pile type/material	Closed-ended steel pipe pile
Length, L (m)	14.6
Outer diameter, D (mm)	406
Installation method	Driven
Set up time, days	28
Loading mode	Compression
$Q_{max\text{-}measured}$ (kN)	1790
Q_m (kN)	1620
Q_s (kN)	Not isolated
Q_b (kN)	Not isolated
API Q_c (kN)	1492
Q_c/Q_m	0.92
UWA Q_c (kN)	3188
Q_c/Q_m	1.97
ICP Q_c (kN)	2827
Q_c/Q_m	1.75
Fugro Q_c (kN)	3760
Q_c/Q_m	2.32
NGI Q_c (kN)	4381
Q_c/Q_m	2.70

Pile ID: 3-2

Load–displacement data

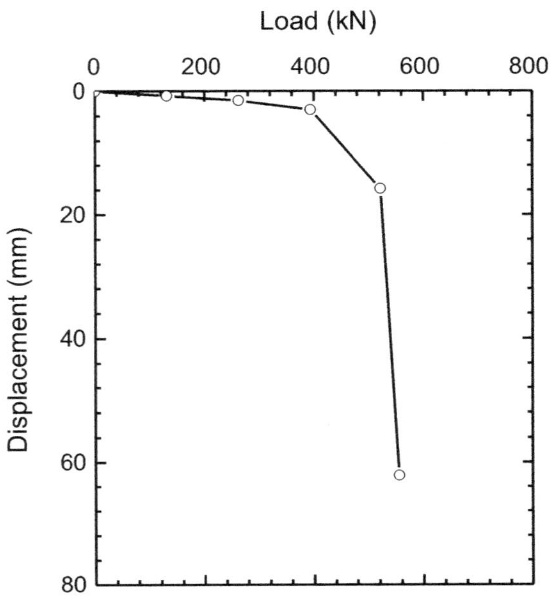

Detail	Description
Pile type/material	Closed-ended steel pipe pile
Length, L (m)	11
Outer diameter, D (mm)	305
Installation method	Driven
Set up time, days	35
Loading mode	Tension
$Q_{max\text{-}measured}$ (kN)	540
Q_m (kN)	540
Q_s (kN)	540
Q_b (kN)	—
API Q_c (kN)	296
Q_c/Q_m	0.55
UWA Q_c (kN)	599
Q_c/Q_m	1.11
ICP Q_c (kN)	554
Q_c/Q_m	1.03
Fugro Q_c (kN)	652
Q_c/Q_m	1.21
NGI Q_c (kN)	1154
Q_c/Q_m	2.14

Pile ID: 3-5

Load–displacement data

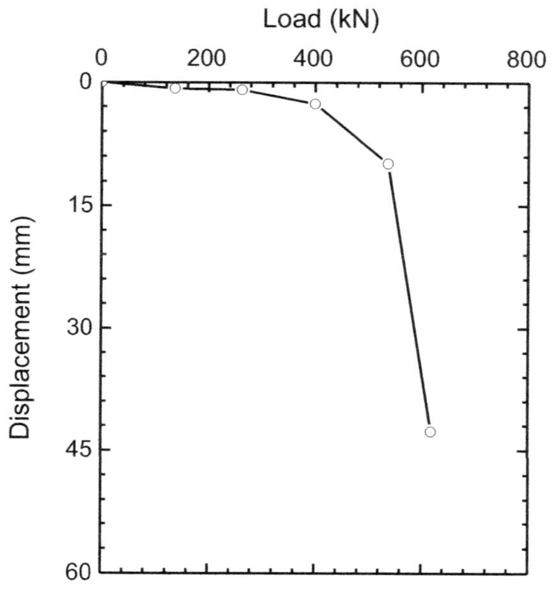

Load (kN)
0 200 400 600 800
Displacement (mm)
0 15 30 45 60

Detail	Description
Pile type/material	Closed-ended steel pipe pile
Length, L (m)	11.1
Outer diameter, D (mm)	305
Installation method	Driven
Set up time, days	27
Loading mode	Tension
$Q_{max\text{-}measured}$ (kN)	610
Q_m (kN)	610
Q_s (kN)	610
Q_b (kN)	—
API Q_c (kN)	353
Q_c/Q_m	0.58
UWA Q_c (kN)	740
Q_c/Q_m	1.21
ICP Q_c (kN)	668
Q_c/Q_m	1.10
Fugro Q_c (kN)	847
Q_c/Q_m	1.39
NGI Q_c (kN)	1367
Q_c/Q_m	2.24

Site ID No. 39

Pile ID: 3-8

Load–displacement data

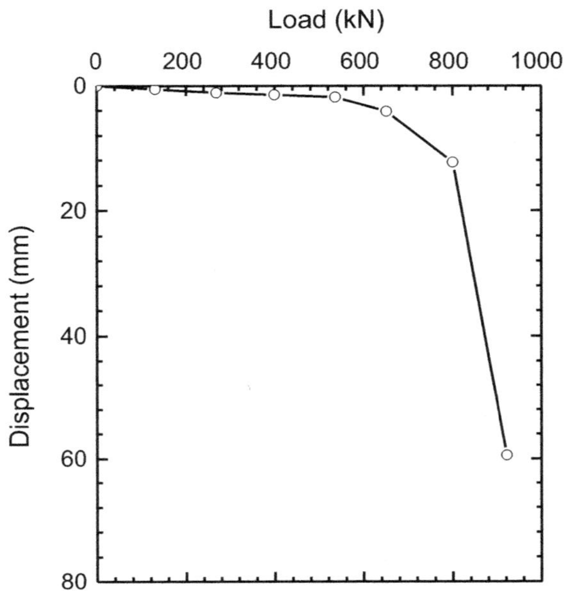

Load (kN)

0 200 400 600 800 1000

Displacement (mm)

0 20 40 60 80

Detail	Description
Pile type/material	Closed-ended steel pipe pile
Length, L (m)	11
Outer diameter, D (mm)	406
Installation method	Driven
Set up time, days	35
Loading mode	Tension
$Q_{max\text{-}measured}$ (kN)	900
Q_m (kN)	900
Q_s (kN)	900
Q_b (kN)	—
API Q_c (kN)	412
Q_c/Q_m	0.46
UWA Q_c (kN)	876
Q_c/Q_m	0.97
ICP Q_c (kN)	776
Q_c/Q_m	0.86
Fugro Q_c (kN)	1036
Q_c/Q_m	1.15
NGI Q_c (kN)	1559
Q_c/Q_m	1.73

Site ID No. 40: Tokyo Bay, Japan.

Ref.: Shioi et al. (1992): Estimation of bearing capacity of steel pipe pile by static loading test and stress wave theory (Trans-Tokyo Bay Highway). Application of stress-wave theory to piles, Balkema, Rotterdam, The Netherlands, 325–330.

Cone penetrometer data

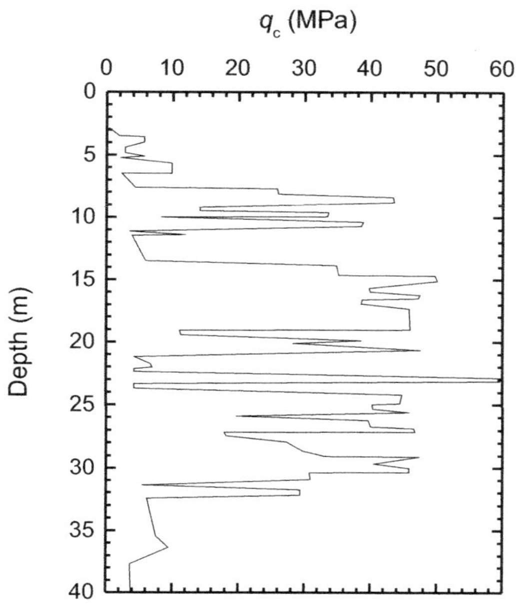

Detail	Description
Site name and location	Tokyo Bay, Japan
Soil type (s)	Loose alluvial sand (to 4 m) over mainly dense sand profile containing thin layers of sandy clay
Water table depth (m)	24.5
Pile type (s)	Open-ended steel pipe pile
Type of cone penetrometer testing	CPT
Number of pile load tests	1
Comments	Only a simplified CPT profile is available. The interface friction angle is 32° measured from ring shear test. Soil unit weight applies default value.

Pile ID: TP

Load–displacement data

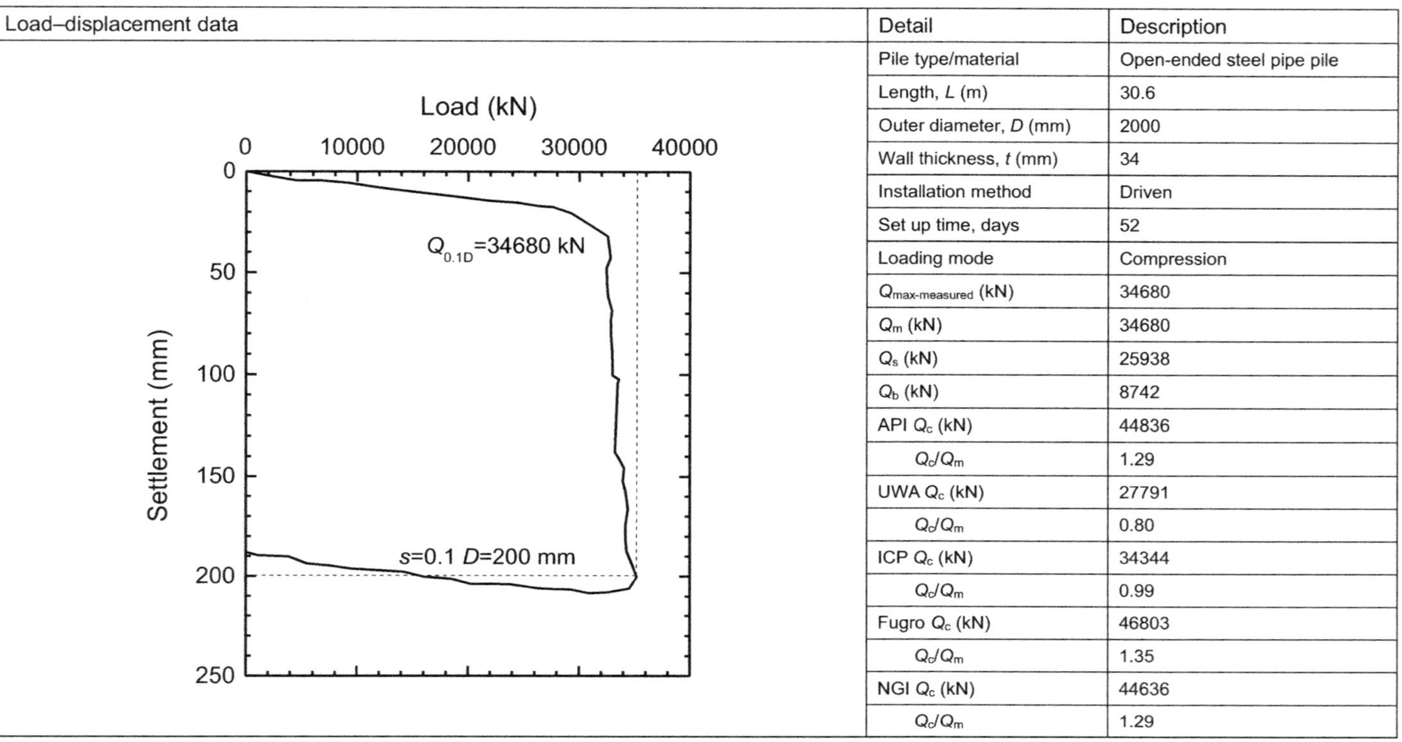

Detail	Description
Pile type/material	Open-ended steel pipe pile
Length, L (m)	30.6
Outer diameter, D (mm)	2000
Wall thickness, t (mm)	34
Installation method	Driven
Set up time, days	52
Loading mode	Compression
$Q_{max\text{-}measured}$ (kN)	34680
Q_m (kN)	34680
Q_s (kN)	25938
Q_b (kN)	8742
API Q_c (kN)	44836
Q_c/Q_m	1.29
UWA Q_c (kN)	27791
Q_c/Q_m	0.80
ICP Q_c (kN)	34344
Q_c/Q_m	0.99
Fugro Q_c (kN)	46803
Q_c/Q_m	1.35
NGI Q_c (kN)	44636
Q_c/Q_m	1.29

Site ID No. 41: Hsin-Ta, Taiwan.

Ref.: Yen et al. (1989): Interpretation of instrumented driven steel pipe piles. Foundation engineering: Current principles and practices, Evanston, ASCE, Reston, VA, 1293–1308.

Cone penetrometer data

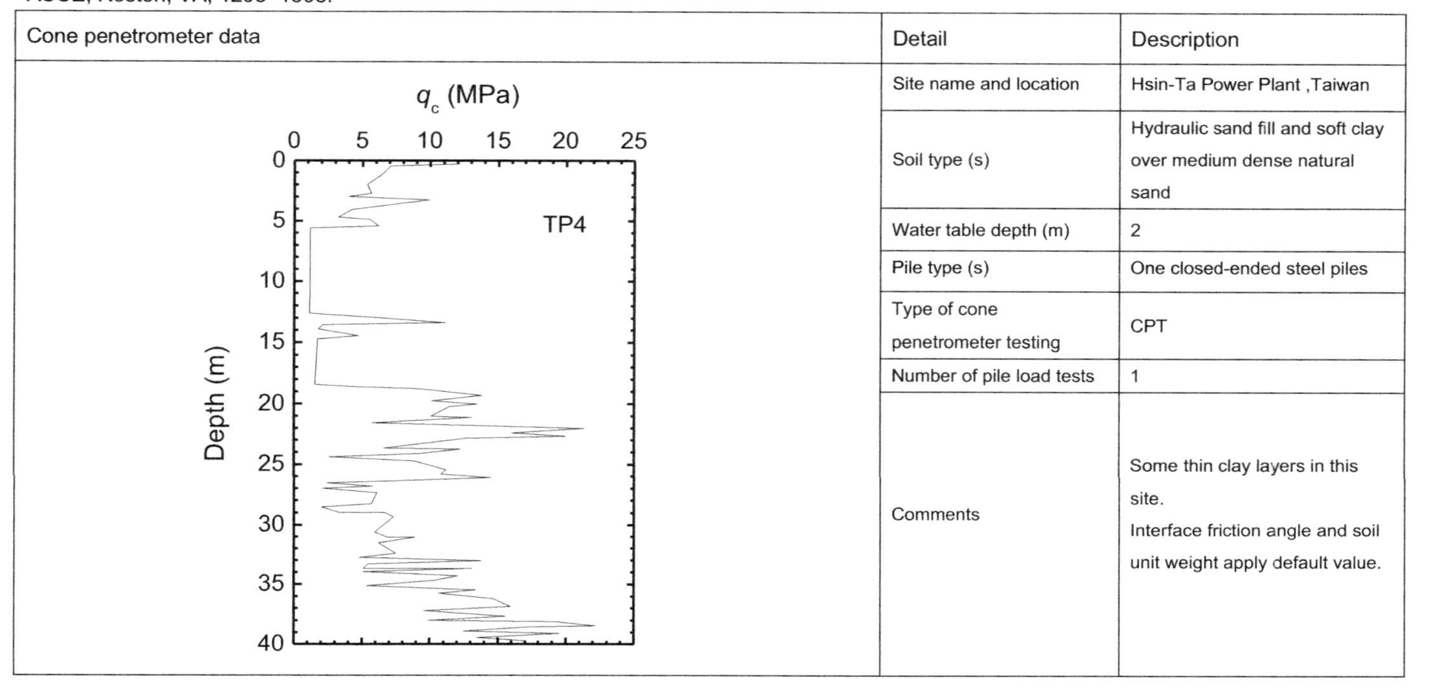

Detail	Description
Site name and location	Hsin-Ta Power Plant ,Taiwan
Soil type (s)	Hydraulic sand fill and soft clay over medium dense natural sand
Water table depth (m)	2
Pile type (s)	One closed-ended steel piles
Type of cone penetrometer testing	CPT
Number of pile load tests	1
Comments	Some thin clay layers in this site. Interface friction angle and soil unit weight apply default value.

Pile ID: TP4

Load–displacement data

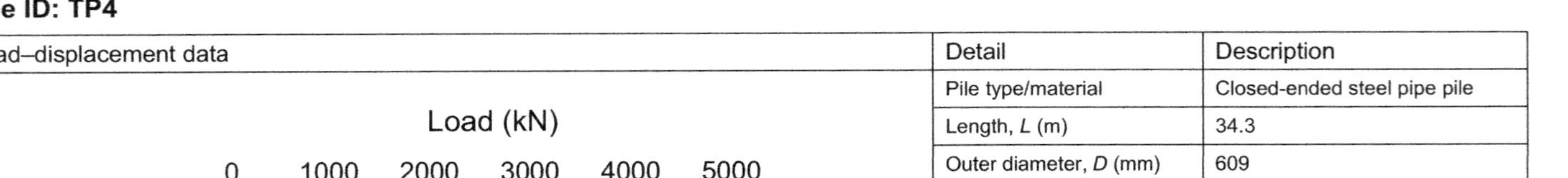

Detail	Description
Pile type/material	Closed-ended steel pipe pile
Length, L (m)	34.3
Outer diameter, D (mm)	609
Installation method	Driven
Set up time, days	33
Loading mode	Compression
$Q_{max\text{-}measured}$ (kN)	4260
Q_m (kN)	4260
Q_s (kN)	3350
Q_b (kN)	910
API Q_c (kN)	3917
Q_c/Q_m	0.92
UWA Q_c (kN)	3454
Q_c/Q_m	0.81
ICP Q_c (kN)	3175
Q_c/Q_m	0.75
Fugro Q_c (kN)	3980
Q_c/Q_m	0.93
NGI Q_c (kN)	4517
Q_c/Q_m	1.06

Site ID No. 42: Hsin-Ta, Taiwan.

Ref.: Yen et al. (1989): Interpretation of instrumented driven steel pipe piles. Foundation engineering: Current principles and practices, Evanston, ASCE, Reston, VA, 1293–1303.

Cone penetrometer data

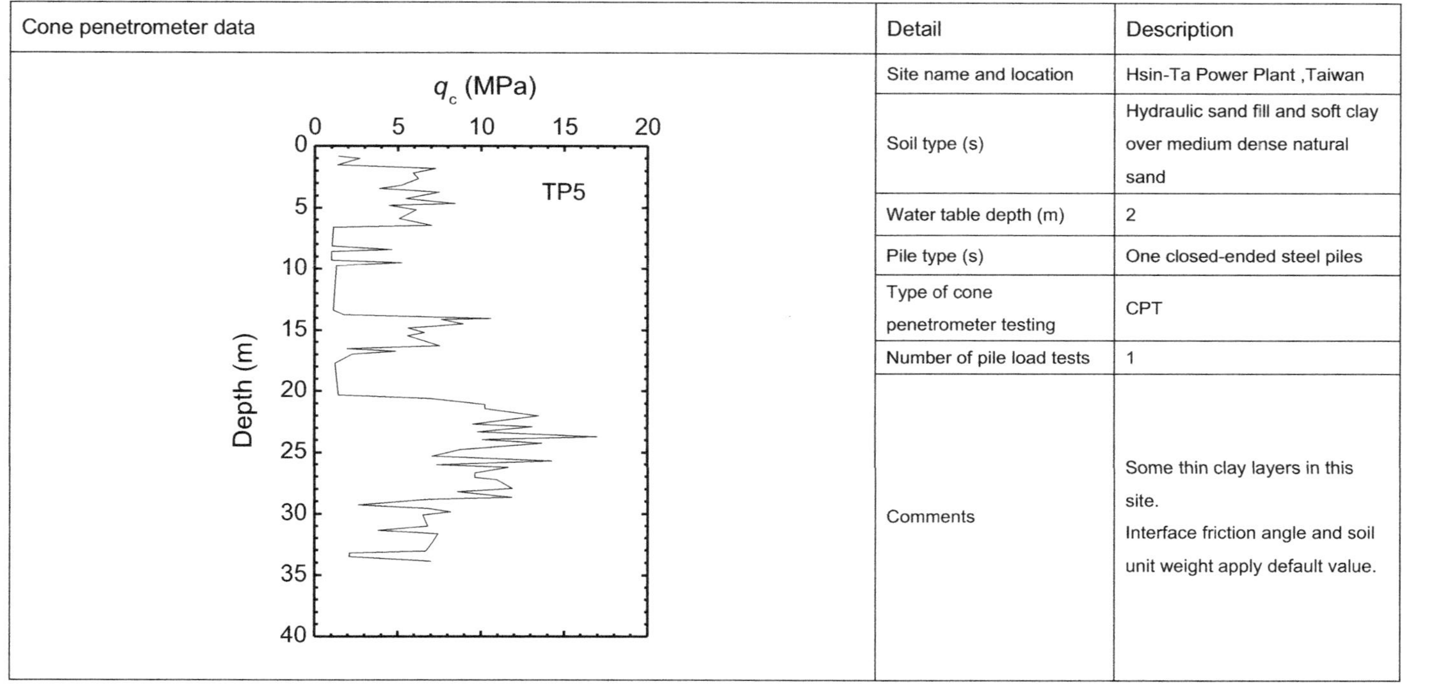

Detail	Description
Site name and location	Hsin-Ta Power Plant ,Taiwan
Soil type (s)	Hydraulic sand fill and soft clay over medium dense natural sand
Water table depth (m)	2
Pile type (s)	One closed-ended steel piles
Type of cone penetrometer testing	CPT
Number of pile load tests	1
Comments	Some thin clay layers in this site. Interface friction angle and soil unit weight apply default value.

Site ID No. 42

Pile ID: TP5

Load–displacement data

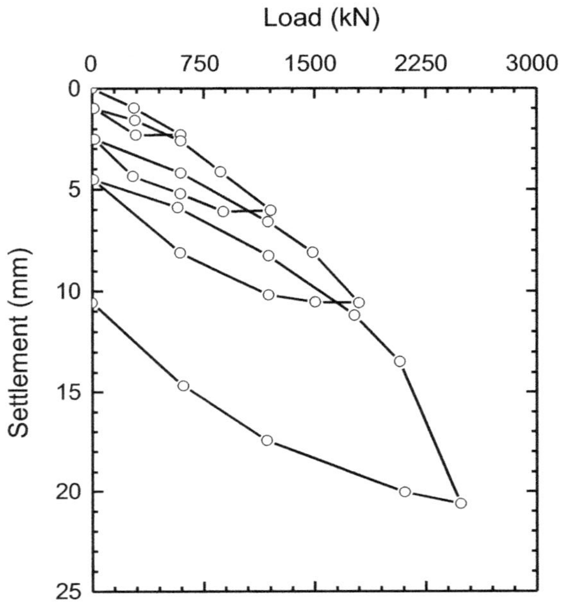
Load (kN)
0
750
1500
2250
3000
Settlement (mm)
0
5
10
15
20
25

Detail	Description
Pile type/material	Closed-ended steel pipe pile
Length, L (m)	34.3
Outer diameter, D (mm)	609
Installation method	Driven
Set up time, days	29
Loading mode	Tension
$Q_{max\text{-}measured}$ (kN)	2450
Q_m (kN)	2630
Q_s (kN)	2630
Q_b (kN)	—
API Q_c (kN)	2855
Q_c/Q_m	1.09
UWA Q_c (kN)	1969
Q_c/Q_m	0.75
ICP Q_c (kN)	2200
Q_c/Q_m	0.84
Fugro Q_c (kN)	1657
Q_c/Q_m	0.63
NGI Q_c (kN)	1911
Q_c/Q_m	0.73

Site ID No. 43: Hsin-Ta, Taiwan.

Ref.: Yen et al. (1989): Interpretation of instrumented driven steel pipe piles. Foundation engineering: Current principles and practices, Evanston, ASCE, Reston, VA, 1293–1308.

Cone penetrometer data

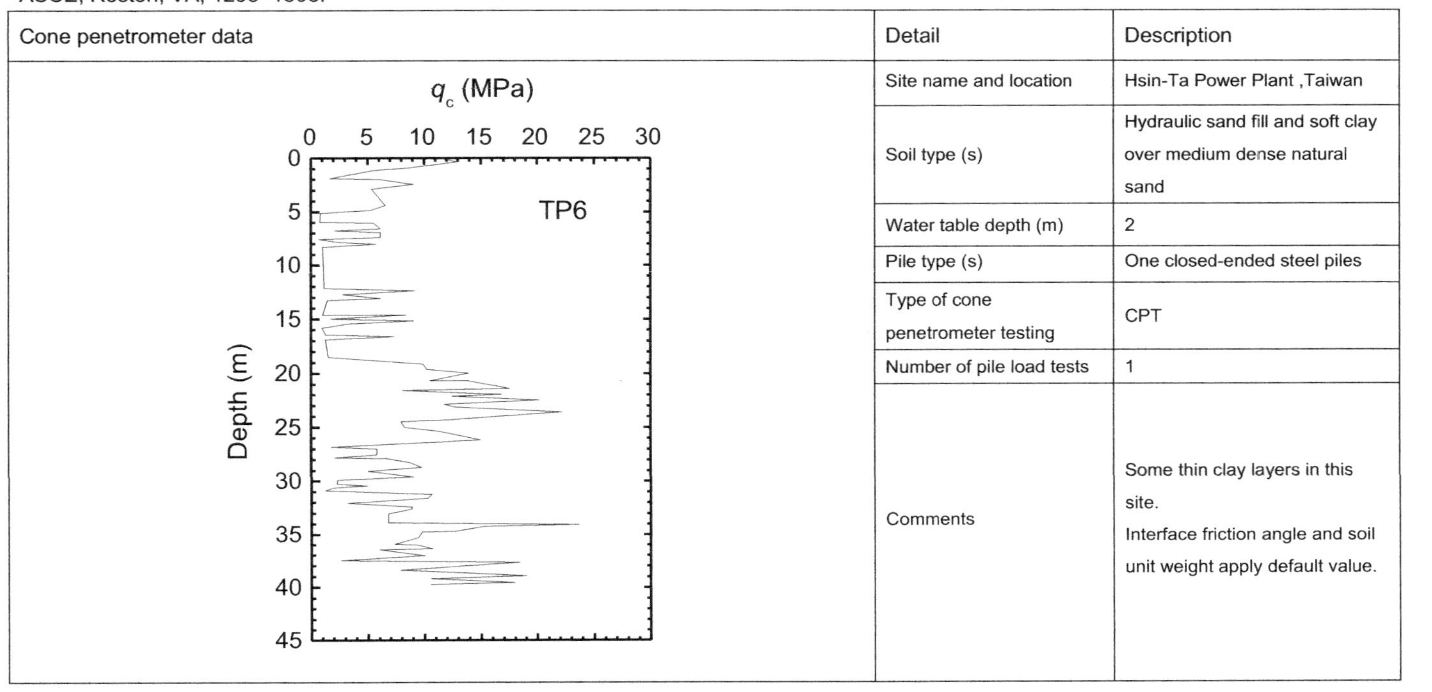

Detail	Description
Site name and location	Hsin-Ta Power Plant ,Taiwan
Soil type (s)	Hydraulic sand fill and soft clay over medium dense natural sand
Water table depth (m)	2
Pile type (s)	One closed-ended steel piles
Type of cone penetrometer testing	CPT
Number of pile load tests	1
Comments	Some thin clay layers in this site. Interface friction angle and soil unit weight apply default value.

Site ID No. 43

Pile ID: TP6

Load–displacement data

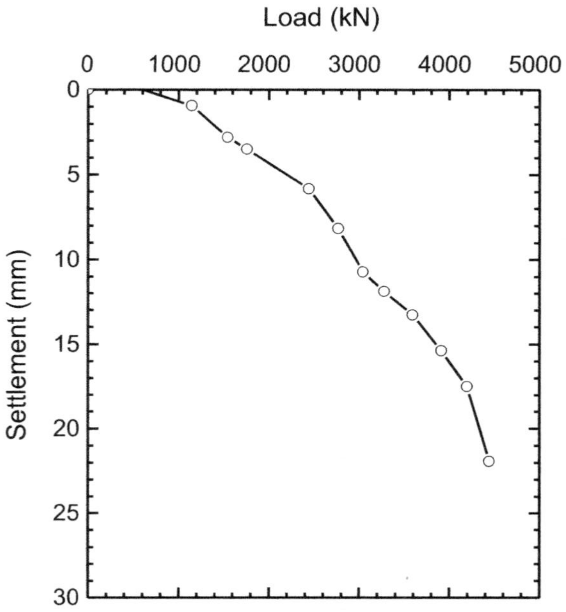

Detail	Description
Pile type/material	Closed-ended steel pipe pile
Length, L (m)	34.3
Outer diameter, D (mm)	609
Installation method	Driven
Set up time, days	30
Loading mode	Compression
$Q_{max\text{-}measured}$ (kN)	4400
Q_m (kN)	4910 (estimated by Hansen's extrapolation)
Q_s (kN)	3810
Q_b (kN)	1100
API Q_c (kN)	2743
Q_c/Q_m	0.56
UWA Q_c (kN)	2791
Q_c/Q_m	0.57
ICP Q_c (kN)	2752
Q_c/Q_m	0.56
Fugro Q_c (kN)	3380
Q_c/Q_m	0.69
NGI Q_c (kN)	2836
Q_c/Q_m	0.58

Site ID No. 44: Drammen, Norway.

Ref.: Tveldt and Fredriksen (2003): N18 Ny motorvegbru I Drammen Prøvebelasting av peler. Conf. Rock and Blasting Geotechnics, Oslo, Norway, 37.1–37.32.

Cone penetrometer data	Detail	Description
q_c (MPa); Depth (m); P16	Site name and location	Drammen river,
	Soil type (s)	Loose to medium dense fine river sand over soft marine clay
	Water table depth (m)	1
	Pile type (s)	Open-ended steel piles
	Type of cone penetrometer testing	CPTU with F_r and pore pressure
	Number of pile load tests	1
	Comments	Pile P16 was driven to three different depths, but only the test at 11 m is accepted, for other two piles were penetrated into thick clay layers below the depth of 15 m. Interface friction angle and soil unit weight apply default value.

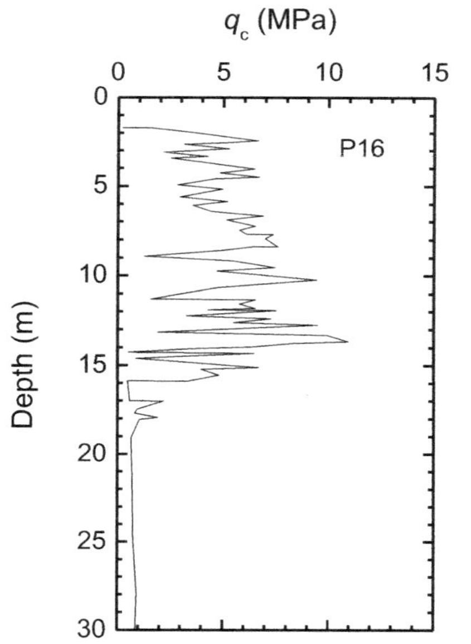

Pile ID: AKSE 16-P1-11

Load–displacement data

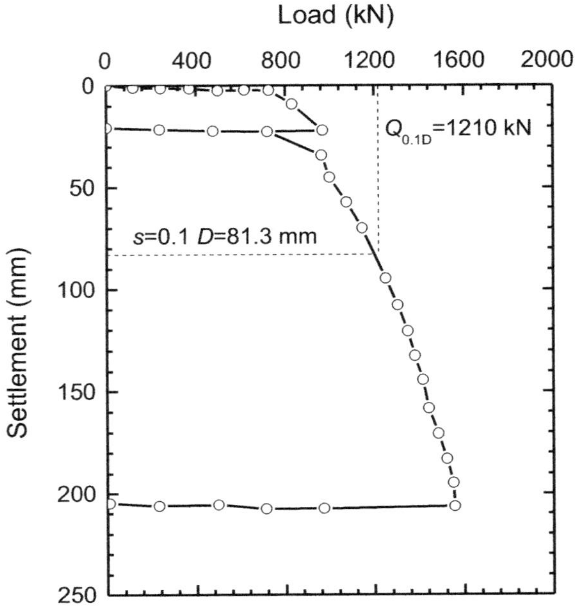

Detail	Description
Pile type/material	Open-ended steel pipe pile
Length, L (m)	11
Outer diameter, D (mm)	813
Wall thickness, t (mm)	12.5
Installation method	Driven
Set up time, days	2
Loading mode	Compression
$Q_{max\text{-}measured}$ (kN)	1600
Q_m (kN)	1210
Q_s (kN)	Not isolated
Q_b (kN)	Not isolated
API Q_c (kN)	1487
Q_c/Q_m	1.23
UWA Q_c (kN)	1007
Q_c/Q_m	0.83
ICP Q_c (kN)	791
Q_c/Q_m	0.65
Fugro Q_c (kN)	2170
Q_c/Q_m	1.79
NGI Q_c (kN)	930
Q_c/Q_m	0.77

Site ID No. 45: Drammen, Norway.

Ref.: Tveldt and Fredriksen (2003): N18 Ny motorvegbru I DrammenPrøvebelasting av peler. Conf. Rock and Blasting Geotechnics, Oslo, Norway, 37.1–37.32.

Cone penetrometer data

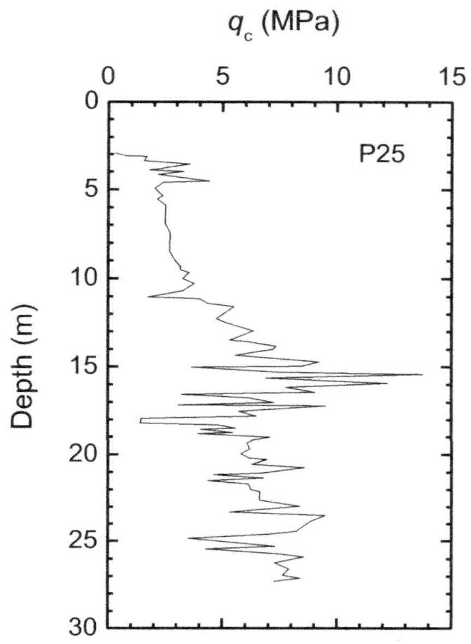

Detail	Description
Site name and location	Drammen river,
Soil type (s)	Loose to medium dense fine river sand over soft marine clay
Water table depth (m)	3
Pile type (s)	Open-ended steel piles
Type of cone penetrometer testing	CPTU with F_r and pore pressure
Number of pile load tests	2
Comments	Pile P25 was driven to two different depths. Interface friction angle and soil unit weight apply default value.

Pile ID: AKSE 25-P2-15

Load–displacement data	Detail	Description
Tabulated values only with no curves given	Pile type/material	Open-ended steel pipe pile
	Length, L (m)	15
	Outer diameter, D (mm)	813
	Wall thickness, t (mm)	12.5
	Installation method	Driven
	Set up time, days	2
	Loading mode	Compression
	$Q_{max\text{-}measured}$ (kN)	2050
	Q_m (kN)	1890
	Q_s (kN)	Not isolated
	Q_b (kN)	Not isolated
	API Q_c (kN)	2233
	Q_c/Q_m	1.18
	UWA Q_c (kN)	1384
	Q_c/Q_m	0.73
	ICP Q_c (kN)	911
	Q_c/Q_m	0.48
	Fugro Q_c (kN)	2510
	Q_c/Q_m	1.33
	NGI Q_c (kN)	1222
	Q_c/Q_m	0.65

Pile ID: AKSE 25-P2-25

Load–displacement data	Detail	Description
Tabulated values only with no curves given	Pile type/material	Open-ended steel pipe pile
	Length, L (m)	25
	Outer diameter, D (mm)	813
	Wall thickness, t (mm)	12.5
	Installation method	Driven
	Set up time, days	2
	Loading mode	Compression
	$Q_{max\text{-}measured}$ (kN)	3280
	Q_m (kN)	2700
	Q_s (kN)	Not isolated
	Q_b (kN)	Not isolated
	API Q_c (kN)	3021
	Q_c/Q_m	1.12
	UWA Q_c (kN)	1842
	Q_c/Q_m	0.68
	ICP Q_c (kN)	1545
	Q_c/Q_m	0.57
	Fugro Q_c (kN)	2605
	Q_c/Q_m	0.96
	NGI Q_c (kN)	2229
	Q_c/Q_m	0.83

Site ID No. 46: Shanghai, China.

Ref.: Pump et al. (1998): Installation and load tests of deep piles in Shanghai alluvium. Proc., 7th Int. Conf. on Piling and Deep Foundations, 1, DFI, Vienna, 31–36.

Cone penetrometer data

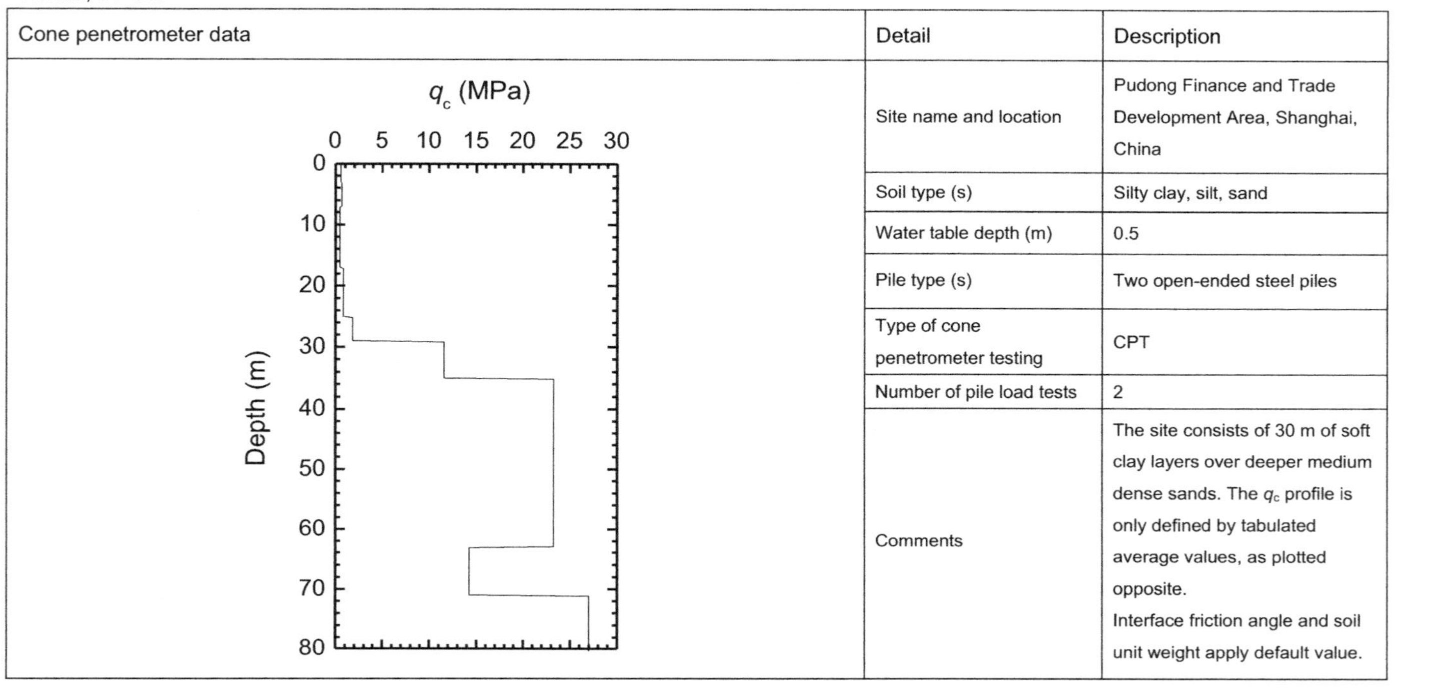

Detail	Description
Site name and location	Pudong Finance and Trade Development Area, Shanghai, China
Soil type (s)	Silty clay, silt, sand
Water table depth (m)	0.5
Pile type (s)	Two open-ended steel piles
Type of cone penetrometer testing	CPT
Number of pile load tests	2
Comments	The site consists of 30 m of soft clay layers over deeper medium dense sands. The q_c profile is only defined by tabulated average values, as plotted opposite. Interface friction angle and soil unit weight apply default value.

Pile ID: ST-1

Load–displacement data

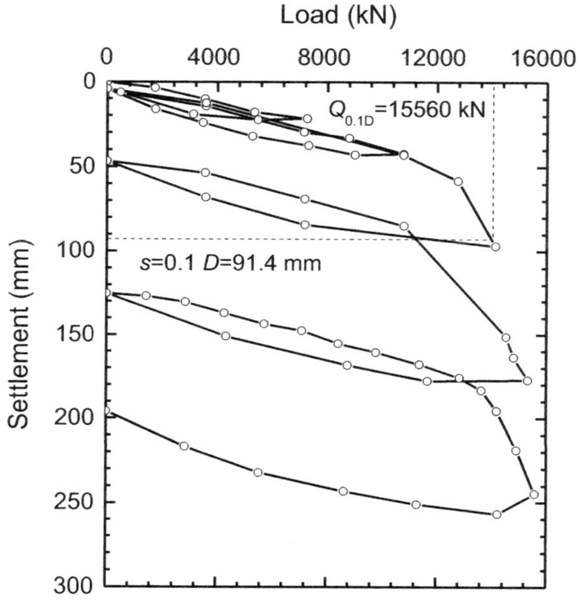

Detail	Description
Pile type/material	Open-ended steel pipe pile
Length, L (m)	79
Outer diameter, D (mm)	914
Wall thickness, t (mm)	20
Installation method	Driven
Set up time, days	23
Loading mode	Compression
$Q_{max\text{-}measured}$ (kN)	16360
Q_m (kN)	15560
Q_s (kN)	11460
Q_b (kN)	4100
API Q_c (kN)	14758
Q_c/Q_m	0.95
UWA Q_c (kN)	13049
Q_c/Q_m	0.84
ICP Q_c (kN)	12968
Q_c/Q_m	0.83
Fugro Q_c (kN)	10760
Q_c/Q_m	0.69
NGI Q_c (kN)	18047
Q_c/Q_m	1.16

Pile ID: ST-2

Load–displacement data

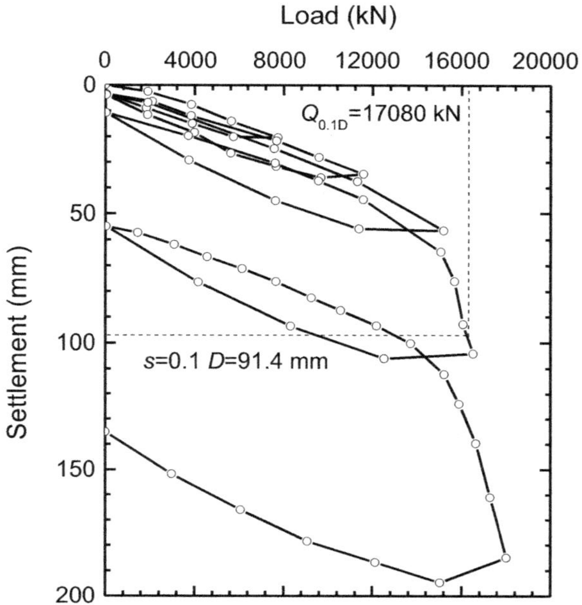

Detail	Description
Pile type/material	Open-ended steel pipe pile
Length, L (m)	79.1
Outer diameter, D (mm)	914
Wall thickness, t (mm)	20
Installation method	Driven
Set up time, days	35
Loading mode	Compression
$Q_{max\text{-}measured}$ (kN)	17820
Q_m (kN)	17080
Q_s (kN)	14570
Q_b (kN)	2510
API Q_c (kN)	14781
Q_c/Q_m	0.87
UWA Q_c (kN)	13067
Q_c/Q_m	0.77
ICP Q_c (kN)	12995
Q_c/Q_m	0.76
Fugro Q_c (kN)	10730
Q_c/Q_m	0.63
NGI Q_c (kN)	18077
Q_c/Q_m	1.06

Site ID No. 47: Cimarron River, USA.

Ref.: Nevels and Snethen (1994): Comparison of settlement predictions for single pile in sand based on penetration test results. Proc. Conf. on Vertical and Horizontal Deformations of Foundations and Embankments, ASCE, Reston, VA, 1028–1038.

Cone penetrometer data

q_c (MPa)
0 5 10 15 20 25 30
Depth (m)
0 5 10 15 20 25
MCPT-1

q_c (MPa)
0 5 10 15 20 25 30
Depth (m)
0 5 10 15 20 25
ECPT-2

Detail	Description
Site name and location	Cimarron River, Northwest Oklahoma, USA
Soil type (s)	Silty shale, mixtures and layers of sand and gravel with silt or clay fines
Water table depth (m)	1
Pile type (s)	One octagonal prestressed concrete pile and one closed-ended steel pipe pile
Type of cone penetrometer testing	MCPT, ECPT
Number of pile load tests	2
Comments	2 MCPT and 2 ECPT soundings were avaiable. While only the two shown opposite are reported by Nevels and Snethan they show good agreement. Interface friction angle and soil unit weight apply default value.

Site ID No. 47

Pile ID: P1

Load–displacement data

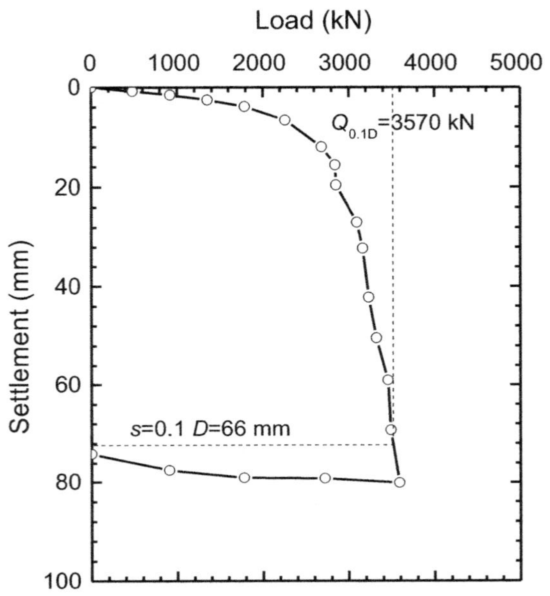
Load (kN)
0
1000
2000
3000
4000
5000
Settlement (mm)
0
20
40
60
80
100
$Q_{0.1D}$=3570 kN
s=0.1 D=66 mm

Detail	Description
Pile type/material	Closed-ended steel pipe pile
Length, L (m)	19
Outer diameter, D (mm)	660
Installation method	Driven
Set up time, days	—
Loading mode	Compression
$Q_{max\text{-}measured}$ (kN)	3580
Q_m (kN)	3570
Q_s (kN)	Not isolated
Q_b (kN)	Not isolated
API Q_c (kN)	3303
Q_c/Q_m	0.93
UWA Q_c (kN)	4622
Q_c/Q_m	1.29
ICP Q_c (kN)	3711
Q_c/Q_m	1.04
Fugro Q_c (kN)	6150
Q_c/Q_m	1.72
NGI Q_c (kN)	6517
Q_c/Q_m	1.83

Pile ID: P2

Load–displacement data

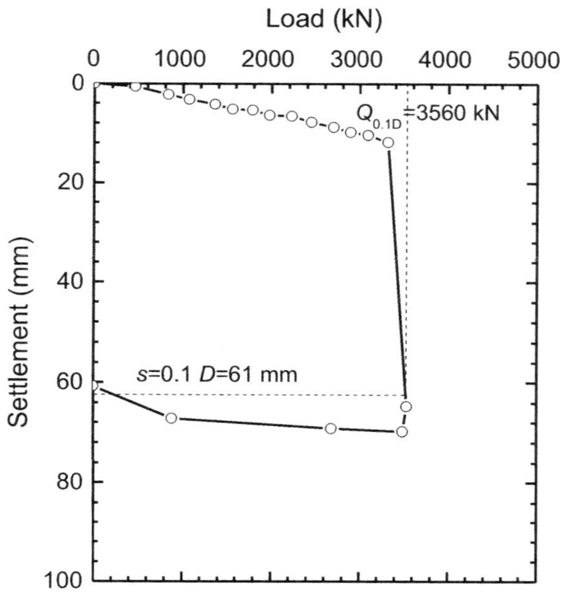

Detail	Description
Pile type/material	Octagonal concrete pile
Length, L (m)	19.5
Outer width, B (mm)	610
Installation method	Driven
Set up time, days	—
Loading mode	Compression
$Q_{max\text{-}measured}$ (kN)	3560
Q_m (kN)	3560
Q_s (kN)	Not isolated
Q_b (kN)	Not isolated
API Q_c (kN)	4290
Q_c/Q_m	1.21
UWA Q_c (kN)	4537
Q_c/Q_m	1.27
ICP Q_c (kN)	3479
Q_c/Q_m	0.98
Fugro Q_c (kN)	6080
Q_c/Q_m	1.71
NGI Q_c (kN)	6848
Q_c/Q_m	1.92

Site ID No. 48: Jonkoping, Sweden.

Ref.: Jendeby et al. (1994): Friction piles in sand-Prediction of bearing capacity and load/displacement curve. Proc. Int. Conf. and Exhibition on Piling and Deep Foundations, DFI, Hawthorn, NJ, 5, DFI 94, 3.6.1–3.6.5.

Cone penetrometer data

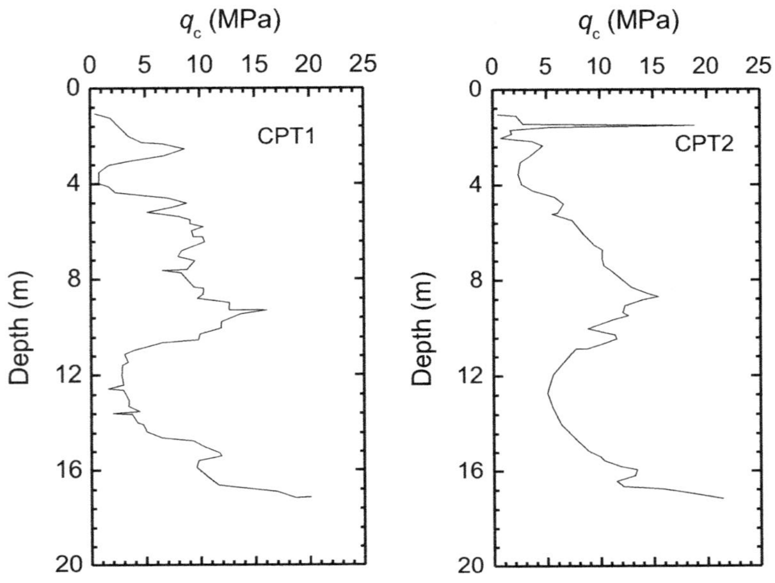

Detail	Description
Site name and location	Jonkoping, southern Sweden
Soil type (s)	Fill, mainly poor graded sand
Water table depth (m)	1.3
Pile type (s)	3 square concrete piles
Type of cone penetrometer testing	CPT
Number of pile load tests	3
Comments	Two CPT profiles are reported that show the same general shapes. An average profile has been adopted for the capacity calculations. Interface friction angle and soil unit weight apply default value.

Pile ID: P23

Load–displacement data

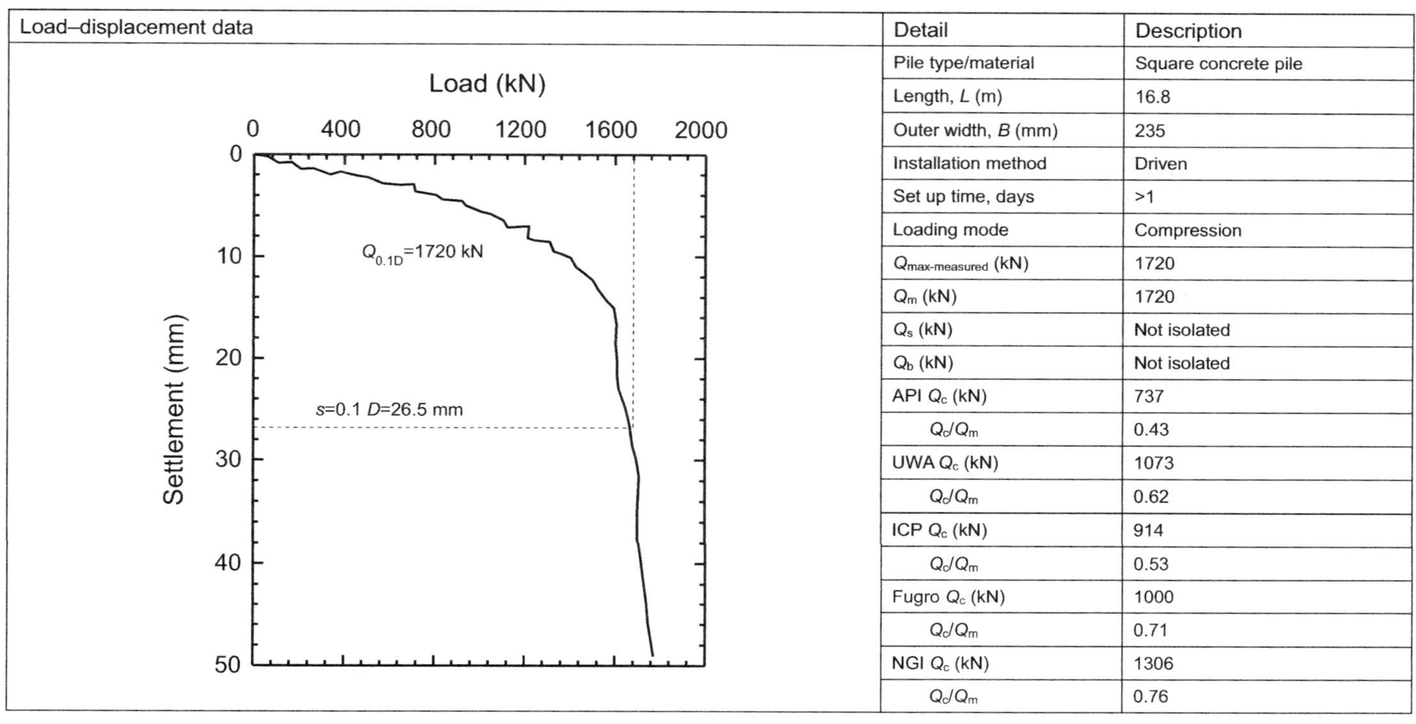

Detail	Description
Pile type/material	Square concrete pile
Length, L (m)	16.8
Outer width, B (mm)	235
Installation method	Driven
Set up time, days	>1
Loading mode	Compression
$Q_{max\text{-}measured}$ (kN)	1720
Q_m (kN)	1720
Q_s (kN)	Not isolated
Q_b (kN)	Not isolated
API Q_c (kN)	737
Q_c/Q_m	0.43
UWA Q_c (kN)	1073
Q_c/Q_m	0.62
ICP Q_c (kN)	914
Q_c/Q_m	0.53
Fugro Q_c (kN)	1000
Q_c/Q_m	0.71
NGI Q_c (kN)	1306
Q_c/Q_m	0.76

Pile ID: P25

Load–displacement data

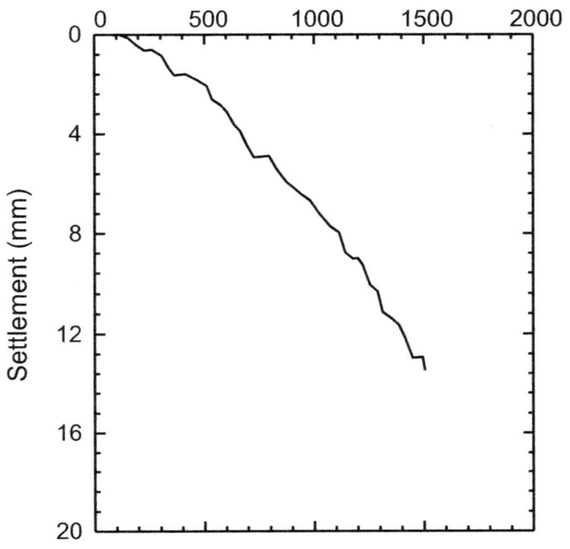

Detail	Description
Pile type/material	Square concrete pile
Length, L (m)	17.8
Outer width, B (mm)	235
Installation method	Driven
Set up time, days	<1
Loading mode	Compression
$Q_{max\text{-}measured}$ (kN)	1500
Q_m (kN)	1650 (tabulated value in the original paper)
Q_s (kN)	Not isolated
Q_b (kN)	Not isolated
API Q_c (kN)	1014
Q_c/Q_m	0.61
UWA Q_c (kN)	1319
Q_c/Q_m	0.80
ICP Q_c (kN)	1111
Q_c/Q_m	0.67
Fugro Q_c (kN)	1170
Q_c/Q_m	0.71
NGI Q_c (kN)	1545
Q_c/Q_m	0.94

Pile ID: P26

Load–displacement data

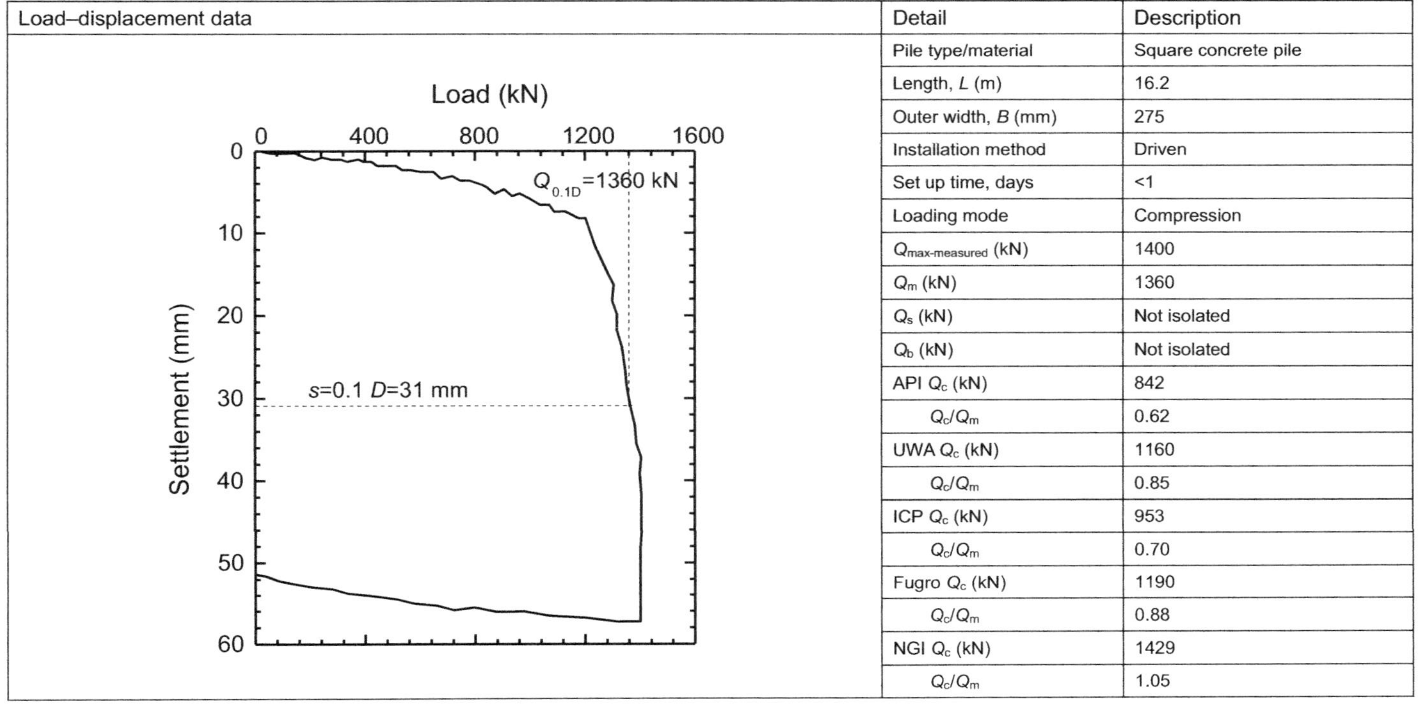

Detail	Description
Pile type/material	Square concrete pile
Length, L (m)	16.2
Outer width, B (mm)	275
Installation method	Driven
Set up time, days	<1
Loading mode	Compression
$Q_{max\text{-}measured}$ (kN)	1400
Q_m (kN)	1360
Q_s (kN)	Not isolated
Q_b (kN)	Not isolated
API Q_c (kN)	842
Q_c/Q_m	0.62
UWA Q_c (kN)	1160
Q_c/Q_m	0.85
ICP Q_c (kN)	953
Q_c/Q_m	0.70
Fugro Q_c (kN)	1190
Q_c/Q_m	0.88
NGI Q_c (kN)	1429
Q_c/Q_m	1.05

Site ID No. 49: Fittja Straits, Sweden.

Ref.: Axelsson (2000): Long term setup of driven piles in sand. (Ph.D. thesis), Royal Institute of Technology, Stockholm, Sweden.

Cone penetrometer data

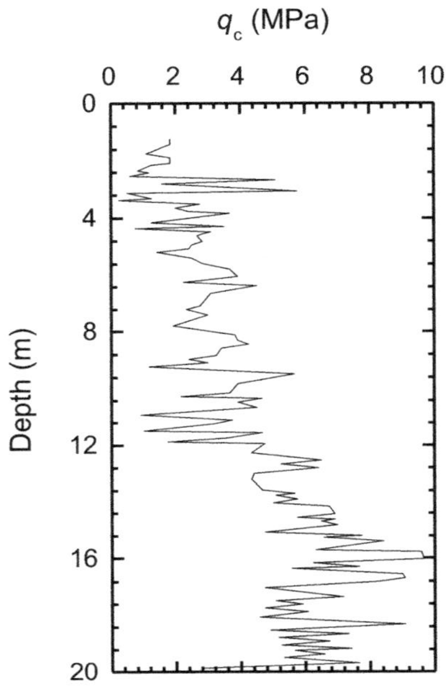

Detail	Description
Site name and location	Fittja Straits, south-west of Stockholm,Sweden
Soil type (s)	Mainly silty sand, with thin clay layers.
Water table depth (m)	2.5
Pile type (s)	Square concrete pile
Type of cone penetrometer testing	CPTU with F_r and pore pressure
Number of pile load tests	2
Comments	All together six load tests, but only two virgin static load tests are accepted, as the others are dynamic tests. Interface friction angle and soil unit weight apply default value.

Pile ID: D-5

Load–displacement data

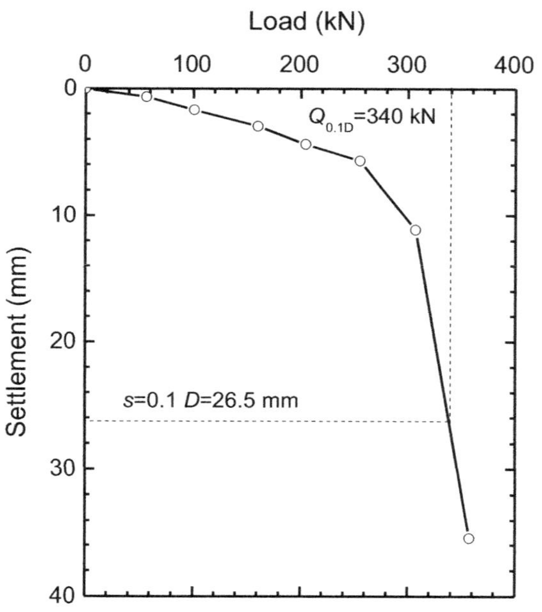

Detail	Description
Pile type/material	Square concrete pile
Length, L (m)	12.8
Outer width, B (mm)	235
Installation method	Driven
Set up time, days	5
Loading mode	Compression
$Q_{max\text{-}measured}$ (kN)	360
Q_m (kN)	340
Q_s (kN)	Not isolated
Q_b (kN)	Not isolated
API Q_c (kN)	388
Q_c/Q_m	1.14
UWA Q_c (kN)	352
Q_c/Q_m	0.94
ICP Q_c (kN)	320
Q_c/Q_m	1.03
Fugro Q_c (kN)	537
Q_c/Q_m	1.58
NGI Q_c (kN)	410
Q_c/Q_m	1.21

Site ID No. 49

Pile ID: D-1*

Load–displacement data

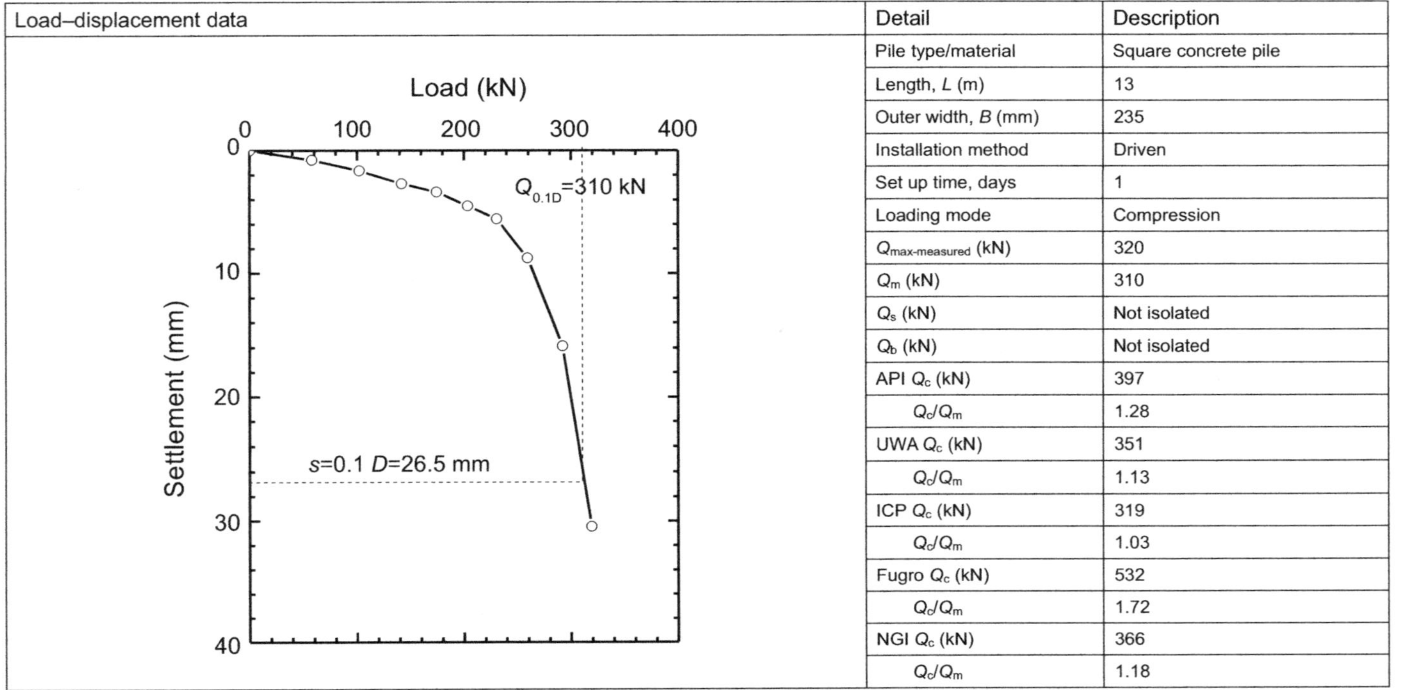

Detail	Description
Pile type/material	Square concrete pile
Length, L (m)	13
Outer width, B (mm)	235
Installation method	Driven
Set up time, days	1
Loading mode	Compression
$Q_{max\text{-}measured}$ (kN)	320
Q_m (kN)	310
Q_s (kN)	Not isolated
Q_b (kN)	Not isolated
API Q_c (kN)	397
Q_c/Q_m	1.28
UWA Q_c (kN)	351
Q_c/Q_m	1.13
ICP Q_c (kN)	319
Q_c/Q_m	1.03
Fugro Q_c (kN)	532
Q_c/Q_m	1.72
NGI Q_c (kN)	366
Q_c/Q_m	1.18

*Pile D-1 was from the same site of Pile D-5 but was not accepted by UWA database. We include this case in ZJU-ICL database as it was re-driven for 0.2 m and considered as a new pile.

Site ID No. 50: Sermide, Italy.

Ref.: Appendino (1981): Interpretation of axial load tests on long piles. Proc., 10th Int. Conf. on Soil Mechanics and Foundation Engineering, Vol. 2, Balkema, Rotterdam, 593–598.

Cone penetrometer data

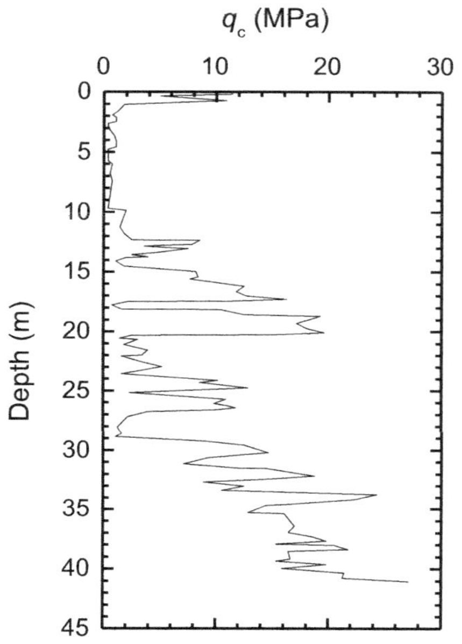

Detail	Description
Site name and location	Sermide, Italy
Soil type (s)	Silty sand, silty clay
Water table depth (m)	0
Pile type (s)	One closed-ended steel pipe piles
Type of cone penetrometer testing	CPT
Number of pile load tests	1
Comments	Three piles all together, but only one pile was loaded to failure. Interface friction angle and soil unit weight apply default value.

Pile ID: S

Load–displacement data

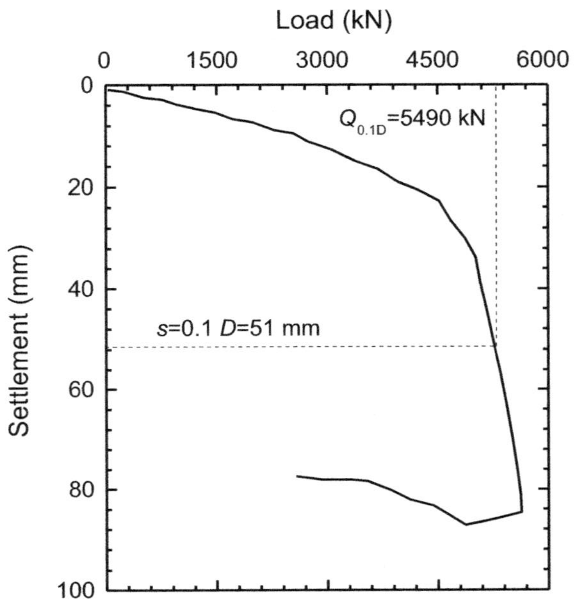

Detail	Description
Pile type/material	Closed-ended steel pipe pile
Length, L (m)	35.9
Outer diameter, D (mm)	508
Installation method	Driven
Set up time, days	—
Loading mode	Compression
$Q_{max\text{-}measured}$ (kN)	5620
Q_m (kN)	5490
Q_s (kN)	3290
Q_b (kN)	2200
API Q_c (kN)	2939
Q_c/Q_m	0.54
UWA Q_c (kN)	4050
Q_c/Q_m	0.74
ICP Q_c (kN)	3606
Q_c/Q_m	0.66
Fugro Q_c (kN)	4190
Q_c/Q_m	0.76
NGI Q_c (kN)	4625
Q_c/Q_m	0.84

Site ID No. 51: Pigeon River, USA.

Ref.: Paik et al. (2003): Behavior of open and closed-ended piles driven into sands. Journal of Geotechnical and Geoenvironmental Engineering, ASCE, 129(4), 296–306.

Cone penetrometer data

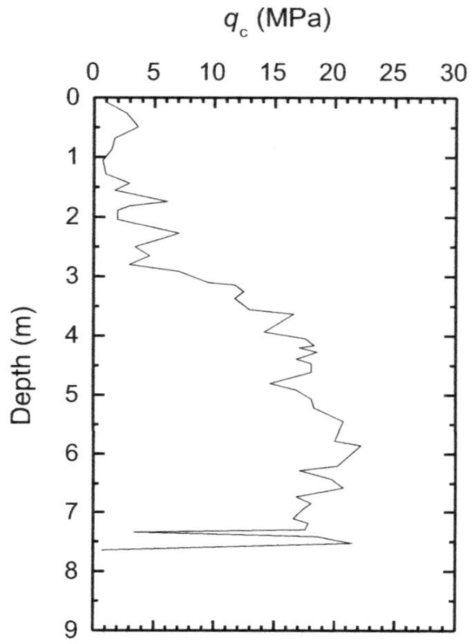

q_c (MPa)

0 5 10 15 20 25 30

Depth (m)

0 1 2 3 4 5 6 7 8 9

Detail	Description
Site name and location	Pigeon River, Lagrange County, Indiana, USA
Soil type (s)	Predominantly gravelly sand beneath clay and silt layers to 3 m depth.
Water table depth (m)	3
Pile type (s)	One closed-ended steel pile
Type of cone penetrometer testing	CPT
Number of pile load tests	1
Comments	The shaft load distributions are reported. Interface friction angle and soil unit weight all apply default value.

Site ID No. 51

Pile ID: 1

Load–displacement data

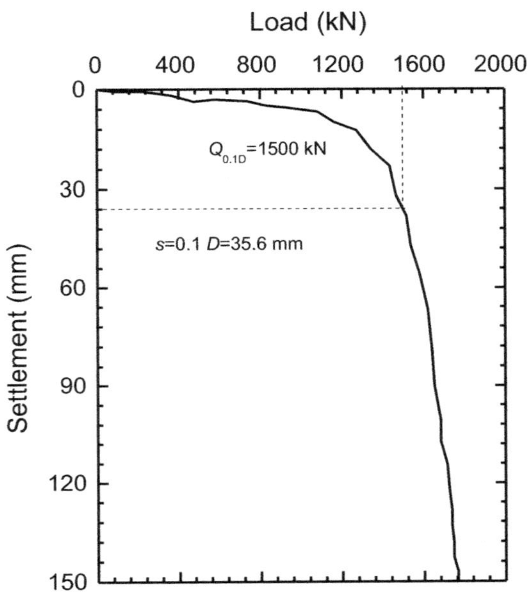

Detail	Description
Pile type/material	Closed-ended steel pipe pile
Length, L (m)	6.9
Outer diameter, D (mm)	356
Installation method	Driven
Set up time, days	4
Loading mode	Compression
$Q_{max\text{-}measured}$ (kN)	1765
Q_m (kN)	1500
Q_s (kN)	633
Q_b (kN)	867
API Q_c (kN)	683
Q_c/Q_m	0.46
UWA Q_c (kN)	2064
Q_c/Q_m	1.38
ICP Q_c (kN)	1773
Q_c/Q_m	1.18
Fugro Q_c (kN)	2230
Q_c/Q_m	1.49
NGI Q_c (kN)	1923
Q_c/Q_m	1.28

Site ID No. 52: Pigeon River, USA.

Ref.: Paik et al. (2003): Behavior of open and closed-ended piles driven into sands. Journal of Geotechnical and Geoenvironmental Engineering, ASCE, 129(4), 296–306.

Cone penetrometer data

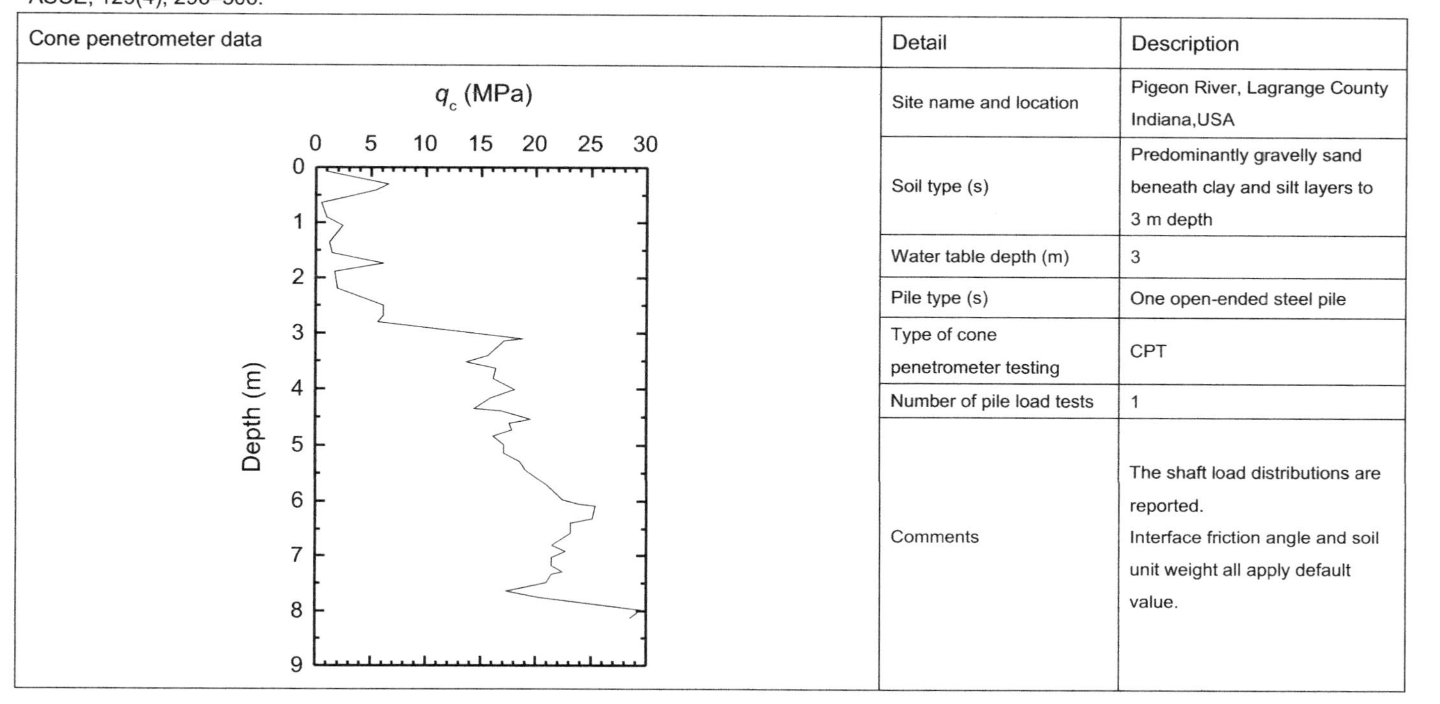

Detail	Description
Site name and location	Pigeon River, Lagrange County Indiana,USA
Soil type (s)	Predominantly gravelly sand beneath clay and silt layers to 3 m depth
Water table depth (m)	3
Pile type (s)	One open-ended steel pile
Type of cone penetrometer testing	CPT
Number of pile load tests	1
Comments	The shaft load distributions are reported. Interface friction angle and soil unit weight all apply default value.

Pile ID: 2

Load–displacement data

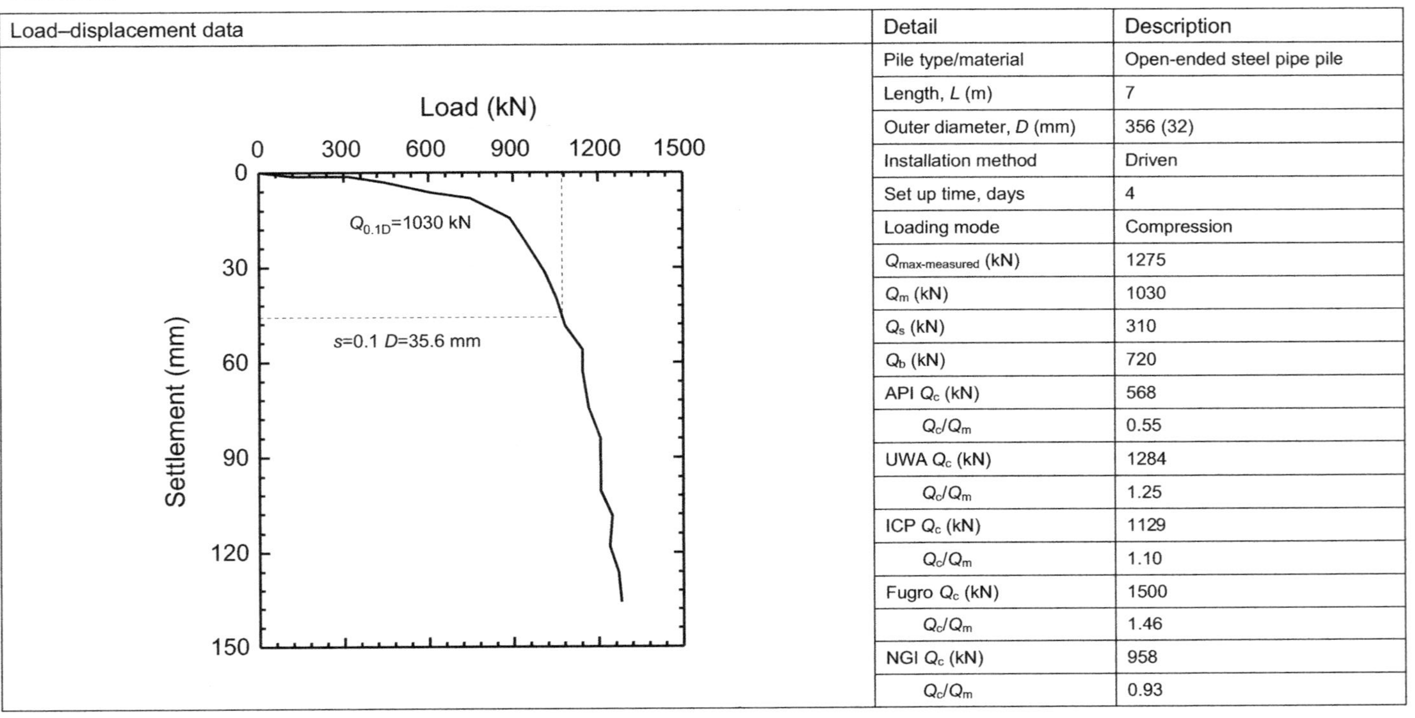

Detail	Description
Pile type/material	Open-ended steel pipe pile
Length, L (m)	7
Outer diameter, D (mm)	356 (32)
Installation method	Driven
Set up time, days	4
Loading mode	Compression
$Q_{max\text{-}measured}$ (kN)	1275
Q_m (kN)	1030
Q_s (kN)	310
Q_b (kN)	720
API Q_c (kN)	568
Q_c/Q_m	0.55
UWA Q_c (kN)	1284
Q_c/Q_m	1.25
ICP Q_c (kN)	1129
Q_c/Q_m	1.10
Fugro Q_c (kN)	1500
Q_c/Q_m	1.46
NGI Q_c (kN)	958
Q_c/Q_m	0.93

References

Altaee, A., Fellenius, B.H., Evgin, E., 1992. Axial load transfer for piles in sand. I. Tests on an instrumented precast pile. Canadian Geotechnical Journal 29 (1), 11–20.

American Petroleum Institute (API), 1993. Recommended Practice for Planning, Designing and Constructing Fixed Offshore Platforms-Working Stress Design, API RP2A, twentieth ed. Washington, DC.

American Petroleum Institute (API), 2000. Recommended Practice for Planning, Designing, and Constructing Fixed Offshore Platforms-Working Stress Design, API RP2A, twenty-first ed. Washington, DC.

American Petroleum Institute (API), 2006. DRAFT Recommended Practice for Planning, Designing, and Constructing Fixed Offshore Platforms-Working Stress Design, API RP2A, twenty-second ed. Washington, DC.

American Petroleum Institute (API), 2014. ANSI/API Recommended Practice 2GEO, first ed. RP2GEO, Washington, DC.

Appendino, M., 1981. Interpretation of axial load tests on long piles. In: Proc., 10th Int. Conf. on Soil Mechanics and Foundation Engineering, vol. 2. Balkema, Rotterdam, pp. 593–598.

Axelsson, G., 2000. Long Term Setup of Driven Piles in Sand (Ph.D thesis). Royal Institute of Technology, Stockholm, Sweden.

Barmpopoulos, I.H., Ho, T.Y.K., Jardine, R.J., Anh-Minh, N., 2009. The large displacement shear characteristics of granular media against concrete and steel interfaces. In: Frost, J.D. (Ed.), Proc. Research Symposium on the Characterization and Behaviour of Interfaces (CBI). Atlanta. IOS Press, Amsterdam, pp. 17–24.

BCP-Committee, 1971. Field tests on piles in sand. Soils and Foundations 11 (2), 29–49.

Beringen, F.L., Windle, D., Van Hooydonk, W.R., 1979. Results of Loading Tests on Driven Piles in Sand. Recent Development in the Design and Construction of Piles. ICE, London. 213–225.

Briaud, J.-L., Tucker, L.M., 1988. Measured and predicted axial response of 98 piles. Journal of Geotechnical Engineering 114 (9), 984–1001.

Briaud, J.-L., Tucker, L.M., Ng, E., 1989a. Axially loaded 5 pile group and single pile in sand. In: Proc., 12th Int. Conf. on Soil Mechanics and Foundation Engineering. Balkema, Rotterdam, pp. 1121–1124.

Briaud, J.-L., Moore, B.H., Mitchell, G.B., 1989b. Analysis of Pile Load Tests at Lock and Dam 26. Foundation Engineering: Current Principles and Practices. Evanston, ACSE, Reston, VA. 925–942.

Bustamante, M., Gianeselli, L., 1982. Pile bearing capacity prediction by means of static penetrometer CPT. In: Proceedings of the 2nd European Symposium on Penetration Testing, pp. 493–500.

Chow, F.C., 1997. Investigations into Displacement Pile Behaviour for Offshore Foundations (Ph.D. thesis). Imperial College London, London.

Clausen, C.J.F., Aas, P.M., Karlsrud, K., 2005. Bearing capacity of driven piles in sand, the NGI approach. In: Proc., Int. Symp. on Frontiers in Offshore Geotechnics. Taylor & Francis, London, pp. 677–681.

CUR, 2001. Bearing Capacity of Steel Pipe Piles Report 2001-8. Centre for Civil Engineering Research and Codes, Gouda, The Netherlands.

de Gijt, J.G., van Dalen, J.H., Middendorp, P., 1995. Comparison of statnamic load test and static load tests at the Rotterdam Harbour. In: First International Statnamic Seminar, Vancouver.

Gavin, K.G., Igoe, D.J.P., Kirwan, L., 2013. The effect of ageing on the axial capacity of piles in sand. Proceedings of the ICE-Geotechnical Engineering 166 (2), 122–130.

Gregersen, O.S., Aas, G., Dibiagio, E., 1973. Load tests on friction piles in loose sand. In: Proc. 8th Int. Conf. on Soil Mechanics and Foundation Engineering, vol. 2.1. pp. 109–117.

Ho, T.Y.K., Jardine, R.J., Anh-Minh, N., 2011. Large-displacement interface shear between steel and granular media. Géotechnique 61 (3), 221–234.

Hölscher, P., 2009. Field Test Rapid Load Testing Waddinxveen. Deltares Factual report, The Netherlands, 95 p.

Jamiolkowski, M.B., Lo Presti, D.F.C., Manassero, M., 2003. Evaluation of Relative Density and Shear Strength of Sands from Cone Penetration Test. Soil Behaviour and Soft Ground Construction. Geotechnical Special Publication, No. 119. ASCE, Reston, VA. 201–238.

Jardine, R.J., 2013. Advanced laboratory testing in research and practice. In: 2nd Bishop Lecture, 18th Int. Conf. on Soil Mechanics and Geotechnical Engineering, Presses des Ponts, pp. 25–55.

Jardine, R.J., Chow, F.C., 1996. New Design Methods for Offshore Piles. Marine Technology Directorate (MTD) Publication 96/103. MTD, London.

Jardine, R.J., Chow, F.C., 2007. Some developments in the design of offshore piles. In: Proc. 6th Int. Conf. on Offshore Site Investigations and Geotechnics. Society for Underwater Technology, London.

Jardine, R.J., Standing, J.R., 2000. Pile Load Testing Performed for HSE Cyclic Loading Study at Dunkirk, France Rep. No. OTO 2000007. Health and Safety Executive, London.

Jardine, R.J., Lehane, B.M., Everton, S.J., 1992. Friction coefficients for piles in sands and silts. In: Proc. 4th Int. Conf. on Offshore Site Investigation and Foundation Behaviour, London, pp. 661–677.

Jardine, R.J., Overy, R.F., Chow, F.C., 1998. Axial capacity of offshore piles in dense North Sea Sand. Journal of Geotechnical and Geoenvironmental Engineering, ASCE 124 (2), 171–178.

Jardine, R.J., Standing, J.R., Chow, F.C., 2006. Some observations of the effects of time on the capacity of piles driven in sand. Géotechnique 56 (4), 227–244.

Jardine, R.J., Chow, F.C., Overy, R., Standing, J.R., 2005. ICP Design Methods for Driven Piles in Sands and Clays. Thomas Telford, London.

Jardine, R.J., Zhu, B.T., Foray, P., Dalton, C.P., 2009. Experimental arrangements for investigation of soil stresses developed around a displacement pile. Soils and Foundations 49 (5), 661–673.

Jardine, R.J., Zhu, B.T., Foray, P., Yang, Z.X., 2013a. Measurement of stresses around closed-ended displacement piles in sand. Géotechnique 63 (1), 1–17.

Jardine, R.J., Zhu, B.T., Foray, P., Yang, Z.X., 2013b. Interpretation of stress measurements made around closed-ended displacement piles in sand. Géotechnique 63 (8), 613–627.

Jendeby, L., Noren, C., Rankka, K., 1994. Friction piles in sand-Prediction of bearing capacity and load/displacement curve. In: Proc. Int. Conf., and Exhibition on Piling and Deep Foundations. DFI, Hawthorn, NJ. 5, DFI 94, 3.6.1-3.6.5.

Karlsrud, K., Jensen, T.G., Wensaas Lied, E.K., Nowacki, F., Simonsen, A.S., 2014. Significant ageing effects for axially loaded piles in sand and clay verified by new field load tests. In: Offshore Technology Conf., Houston. http://dx.doi.org/10.4043/25197-MS.

Kolk, H.J., Baaijens, A.E., Sender, M., 2005a. Design criteria for pipe piles in silica sands. In: Proc. Int. Symp. on Frontiers in Offshore Geotechnics. Taylor & Francis, London, pp. 711–716.

Kolk, H.J., Baaijens, A.E., Vergobi, P., 2005b. Results of axial load tests on pipe piles in very dense sands: the EURIPIDES JIP. In: Proc. Int. Symp. on Frontiers in Offshore Geomechanics, ISFOG. Taylor & Francis, London, pp. 661–667.

Komurka, V.E., Grauvogl-Graham, J.L., 2010. Pile Test Program Report: Lafayette Bridge Replacement Report No. 09019 submitted to Minnesota Department of Transportation. Wagner Komurka Geotechnical Group, Inc., Cedarburg, WI. 844 p.

Lehane, B.M., Jardine, R.J., 1994. Shaft capacity of driven piles in sand: a new design approach. In: Proc. VII Int. Conf. on Behaviour of Offshore Structures, Boston, vol. 1. pp. 23–36.

Lehane, B.M., Schneider, J.A., Xu, X., 2005a. A Review of Design Methods for Offshore Driven Piles in Siliceous Sand UWA Rep. No. GEO 05358. The University of Western Australia, Perth, Australia.

Lehane, B.M., Schneider, J.A., Xu, X., 2005b. The UWA-05 method for prediction of axial capacity of driven piles in sand. In: Gourvenec, S., Cassidy, M. (Eds.), ISFOG 2005: Proc. 1st Int. Symp. On Frontiers in Offshore Geotechnics, University of Western Australia, Perth, 19–21 September 2005. Taylor & Francis, London, pp. 683–689.

Lehane, B.M., Jardine, R.J., Bond, A.J., Frank, R., 1993. Mechanisms of shaft friction in sand from instrumented pile tests. Journal of Geotech Engineering Division, ASCE 119 (1), 19–35.

Mayne, P.W., 2013. FHWA Deep Foundation Load Test Database (DFLTD). Private communication.

Mayne, P.W., Elhakim, A., 2002. Axial pile response evaluation by geophysical piezocone tests. In: Proc. 9th International Conf. On Piling and Deep Foundations, Nice, June 3–5.

Merritt, A., Schroeder, F., Jardine, R.J., Stuyts, B., Cathie, D., Cleverly, W., 2012. Development of pile design methodology for an offshore wind farm in the North Sea. In: Proc. 7th Offshore Site Investigation and Geotechnics: Integrated Geotechnologies-Present and Future. SUT, London.

Naesgaard, E., Amini, A., Uthayakumar, U.M., Fellenius, B.H., 2012. Long Piles in Thick Lacustrine and Deltaic Deposits. Two Bridge Foundation Case Histories. Geotechnical Special Publication, No. 227. ASCE. 404–421.

Nevels, J.B.J., Snethen, D.R., 1994. Comparison of settlement predictions for single pile in sand based on penetration test results. In: Proc., Conf. on Vertical and Horizontal Deformations of Foundations and Embankments. ASCE, Reston, VA, pp. 1028–1038.

Niazi, F.S., 2014. static Axial Pile Foundation Response Using Seismic Piezocone Data (Ph.D thesis). School of Civil and Environmental Engineering, Georgia Institute of Technology, Atlanta, Georgia.

Overy, R., 2007. The use of ICP design methods for the foundations of nine platforms installed in the U.K. North Sea. In: Proc. 6th Int. Conf. on Offshore Site Investigations and Geotechnics. Society for Underwater Technology, London.

Paik, K., Salgado, R., Lee, J., Kim, B., 2003. Behavior of open and closed-ended piles driven into sands. Journal of Geotechnical and Geoenvironmental Engineering, ASCE 129 (4), 296–306.

Pando, M., Filz, G., Ealy, C., Hoppe, E., 2003. Axial and lateral load performance of two composite piles and one prestressed concrete pile. In: TRB 2003 Annual Meeting.

Pump, W., Korista, S., Scott, J., 1998. Installation & load tests of deep piles in Shanghai alluvium. In: Proc. 7th Int. Conf. on Piling and Deep Foundations, vol. 1. DFI, Vienna, pp. 31–36.

Rimoy, S.P., 2013. Ageing and Axial Cyclic Loading Studies of Displacement Piles in Sands (Ph.D dissertation). Imperial College, London.

Rimoy, S.P., Silva, R., Jardine, R.J., Foray, P., Yang, Z.X., Zhu, B.T., Tsuha, C.H.C., 2015. Field and model investigations into the influence of age on axial capacity of displacement piles in silica sands. Géotechnique 65 (7), 576–589.

Rücker, W., Karabeliov, K., Cuéllar, P., Baeßler, M., Georgi, S., 2013. Großversuche an Rammpfählen zur Ermittlung der Tragfähigkeit unter zyklischer Belastung und Standzeit. Geotechnik 36 (2), 77–89.

Schneider, J.A., Xu, X., Lehane, B.M., 2008. Database assessment of CPT-based design methods for axial capacity of driven piles in siliceous sands. Journal of Geotechnical and Geoenvironmental Engineering, ASCE 134 (9), 1227–1244.

Shioi, Y., Yoshida, O., Meta, T., Homma, M., 1992. Estimation of Bearing Capacity of Steel Pipe Pile by Static Loading Test and Stress Wave Theory(Trans-Tokyo Bay Highway). Application of Stress-wave Theory to Piles. Balkema, Rotterdam, The Netherlands, pp. 325–330.

Titi, H., Abu-Farsakh, M., 1999. Evaluation of Bearing Capacity of Piles from Cone Penetration Tests Data. Louisiana Transportation Research Centre, Baton Rouge, LA.

Tsuha, C.H.C., 2012. Companhia Siderúrgica Do Atlântico. Rio de Janeiro. Private communication.

Tsuha, C.H.C., Foray, P.Y., Jardine, R.J., Yang, Z.X., Matias, S., Rimoy, S., 2012. Behaviour of displacement piles in sand under cyclic axial loading. Soils and Foundations 52 (3), 393–410.

Tveldt, G., Fredriksen, F., 2003. N18 Ny motorvegbru I Drammen Prøvebelasting av peler. In: Conf., Rock and Blasting Geotechnics, Oslo, NO, pp. 37.1–37.32.

Vesic, A.S., 1970. Tests on instrumented piles, Ogeechee River site. Journal of the Soil Mechanics and Foundations Division 96 (2), 561–584.

White, D.J., Lehane, B.M., 2004. Friction fatigue on displacement piles in sand. Géotechnique 54 (10), 645–658.

Williams, R.E., Chow, F.C., Jardine, R.J., 1997. Unexpected behaviour of large diameter tubular steel piles. In: Proc. Int. Conf. on Foundation Failures. IES, NTU, NUS and Inst. Structural Engineers, Singapore, pp. 363–378.

Xu, X., 2006. Investigation of the End Bearing Performance of Displacement Piles in Sand (Ph.D dissertation). The University of Western Australia, Perth, Australia.

Yang, Z.X., Jardine, R.J., Zhu, B.T., Foray, P., Tsuha, C.H.C., 2010. Sand grain crushing and interface shearing during displacement pile installation in sand. Géotechnique 60 (6), 469–482.

Yang, Z.X., Jardine, R.J., Zhu, B.T., Rimoy, S., 2014. Stresses developed around displacement piles penetration in sand. Journal of Geotechnical and Geoenvironmental Engineering, ASCE 140 (3), 04013027–4013031.

Yang, Z.X., Guo, W.B., Zha, F.S., Jardine, R.J., Xu, C.J., Cai, Y.Q., 2015a. Field behaviour of driven pre-stressed high-strength concrete piles in sandy soils. Journal of Geotechnical and Geoenvironmental Engineering, ASCE 141 (6), 04015020.

Yang, Z.X., Jardine, R.J., Guo, W.B., Chow, F., 2015b. A new and openly accessible database of tests on piles driven in sands. Géotechnique Letters 5 (1), 12–20.

Yen, T.L., Chin, C.T., Wang, R.F., 1989. Interpretation of Instrumented Driven Steel Pipe Piles. Foundation Engineering: Current Principles and Practices. Evanston, ASCE, Reston, VA. 1293–1308.

Zhang, C., Yang, Z.X., Nguyen, G.D., Jardine, R.J., Einav, I., 2014. Theoretical breakage mechanics and experimental assessment of stresses surrounding piles penetrating into dense silica sand. Géotechnique Letters 4, 11–16.

Index

Note: Page numbers followed by "f" and "t" indicate figures and tables respectively.

S

T

U

W

Z

Printed in the United States
By Bookmasters